Predictive Modeling and Risk Assessment

ISEKI-FOOD SERIES

Series Editor: Kristberg Kristbergsson, University of Iceland
Reykjavík, Iceland

Rui Costa • Kristberg Kristbergsson
Editors

Predictive Modeling and Risk Assessment

 Springer

Editors
Rui Costa
College of Agriculture of the Polytechnic
   Institute of Coimbra
Food Science and Technology Department
Bencanta
3040-316 Coimbra
Portugal

Kristberg Kristbergsson
Department of Food Science
Faculty of Food Science and Nutrition
University of Iceland, Reykjavik
Iceland

Series Editor
Kristberg Kristbergsson
Department of Food Science
Faculty of Food Science and Nutrition
University of Iceland, Reykjavik
Iceland

ISBN: 978-1-4419-4135-0   e-ISBN: 978-0-387-68776-6
DOI: 10.1007/978-0-387-68776-6

Series Preface

The single most important task of food scientists and the food industry as a whole is to ensure the safety of foods supplied to consumers. Recent trends in global food production, distribution and preparation call for increased emphasis on hygienic practices at all levels and for increased research in food safety in order to ensure a safer global food supply. The ISEKI-Food book series is a collection of books where various aspects of food safety and environmental issues are introduced and reviewed by scientists specializing in the field. In all of the books a special emphasis was placed on including case studies applicable to each specific topic. The books are intended for graduate students and senior level undergraduate students as well as professionals and researchers interested in food safety and environmental issues applicable to food safety.

The idea and planning of the books originates from two working groups in the European thematic network "ISEKI-Food" an acronym for "Integrating Safety and Environmental Knowledge In to Food Studies". Participants in the ISEKI-Food network come from 29 countries in Europe and most of the institutes and universities involved with Food Science education at the university level are represented. Some international companies and non teaching institutions have also participated in the program. The ISEKI-Food network is coordinated by Professor Cristina Silva at The Catholic University of Portugal, College of Biotechnology (Escola) in Porto. The program has a web site at: http://www.esb.ucp.pt/iseki/. The main objectives of ISEKI-Food have been to improve the harmonization of studies in food science and engineering in Europe and to develop and adapt food science curricula emphasizing the inclusion of safety and environmental topics. The ISEKI-Food network started on October 1st in 2002, and has recently been approved for funding by the EU for renewal as ISEKI-Food 2 for another three years. ISEKI has its roots in an EU funded network formed in 1998 called Food Net where the emphasis was on casting a light on the different Food Science programs available at the various universities and technical institutions throughout Europe. The work of the ISEKI-Food network was organized into five different working groups with specific task all aiming to fulfill the main objectives of the network.

The first four volumes in the ISEKI-Food book series come from WG2 coordinated by Gerhard Schleining at Boku University in Austria and the undersigned. The main task of the WG2 was to develop and collect materials and methods for teaching of safety and environmental topics in the food science and engineering curricula. The first volume is devoted to Food Safety in general with a practical and a case study approach. The book is composed of fourteen chapters which were organized into three sections on preservation and protection; benefits and risk of microorganisms and process safety. All of these issues have received high public interest in recent years and will continue to be in the focus of consumers and regulatory personnel for years to come. The second volume in the series is devoted to the control of air pollution and treatment of odors in the food industry. The book is divided into eight chapters devoted to defining the problem, recent advances in analysis and methods for prevention and treatment of odors. The topic should be of special interest to industry personnel and researchers due to recent and upcoming regulations by the European Union on air pollution from food processes. Other countries will likely follow suit with more strict regulations on the level of odors permitted to enter the environment from food processing operations. The third volume in the series is devoted to utilization and treatment of waste in the food industry. Emphasis is placed on sustainability of food sources and how waste can be turned into by products rather than pollution or land fills. The Book is composed of 15 chapters starting off with an introduction of problems related to the treatment of waste, and an introduction to the ISO 14001 standard used for improving and maintaining environmental management systems. The book then continues to describe the treatment and utilization of both liquid and solid waste with case studies from many different food processes. The last book from WG2 is on predictive modeling and risk assessment in food products and processes. Mathematical modeling of heat and mass transfer as well as reaction kinetics is introduced. This is followed by a discussion of the stoichiometry of migration in food packaging, as well as the fate of antibiotics and environmental pollutants in the food chain using mathematical modeling and case study samples for clarification.

Volumes five and six come from work in WG5 coordinated by Margarida Vieira at the University of Algarve in Portugal and Roland Verhé at Gent University in Belgium. The main objective of the group was to collect and develop materials for teaching food safety related topics at the laboratory and pilot plant level using practical experimentation. Volume five is a practical guide to experiments in unit operations and processing of foods. It is composed of twenty concise chapters each describing different food processing experiments outlining theory, equipment, procedures, applicable calculations and questions for the students or trainee followed by references. The book is intended to be a practical guide for the teaching of food processing and engineering principles. The final volume in the ISEKI-Food book series is a collection of case studies in food safety and environmental health. It is intended to be a reference for introducing case studies into traditional lecture based safety courses as well as being a basis for problem based learning. The book consists of thirteen chapters containing case studies that may be used, individually or in a series, to discuss a range of food safety issues. For convenience the book was divided into three main sections with the first devoted to case studies, in a more general framework with a number of specific issues in safety and health ranging from acrylamide and nitrates to Botulism and Listeriosis. The second section is devoted to some well known outbreaks related to food intake in different countries. The final section of the book takes on food safety from the perspective of the researcher. Cases are based around experimental data and examine the importance of experimental planning, design and analysis.

The ISEKI-Food books series draws on expertise form close to a hundred universities and research institutions all over Europe. It is the hope of the authors, editors, coordinators and participants in the ISEKI network that the books will be useful to students and colleagues to further there understanding of food safety and environmental issues.

March, 2008   Kristberg Kristbergsson

Series Acknowledgments

ISEKI-Food is a thematic network on food studies, funded by the European Union as project N° 55792-CP-3-00-1-FR-ERASMUS-ETN. It is a part of the EU program in the field of higher education called ERASMUS, which is the higher education action of SOCRATES II program of the EU.

Preface

Mathematical modeling is used in Food Science to predict an outcome when information is incomplete where extrapolation is based on similar conditions. The models are important for all who are interested in preventing risks from hazardous or health threatening situations, for example the growth of certain microorganisms. Models are used to ensure that a specific food process or treatment of a raw material will provide safe and healthy foods. Modeling is particularly useful in risk assessment, to simulate conditions for shelf-life estimation and for the evaluation of packaging materials.

The purpose of this book is to present to food science students, food scientists and food science professionals an overview of available mathematical models in food science and related subjects of particular interest to foods with respect to consumer well being. The book does not intend to be a comprehensive reference for research purposes but will rather illustrate some of the research being conducted to day. The book is a collection of applications that have been developed in diverse fields like pharmacy, agriculture, environmental- and materials science but all apply to foods.

Predictive modeling in food science is typically related to modeling the processing and preservation of food materials in order to achieve or improve food quality and food safety. An example is the modeling of food sterilization by heat treatment which involves both modeling of heat transfer (transport phenomenon) and decay of microorganisms (reaction kinetics). Risk assessment uses models of these phenomena and statistical tools like the Monte Carlo simulation, in order to evaluate the risk of intake of undesirable substances in a population.

The book is divided into three main sections, an introduction, a second section on processing, distribution and consumption of foods and finally there is a section devoted to the fate of pollutants and contaminants in the environment that are significantly related to foods and food processing.

The book starts with an introduction to the basis of most of the mathematical modeling used in food science which are heat and mass transfer theory, the phenomena that are the basis for modeling food processing and food preservation techniques and secondly on reaction kinetics, used to predict molecular activities or growth or decay of microorganisms. This is followed by a general chapter on the quantitative approach to risk assessment in general.

The second section of the book contains six chapters devoted to the modeling of some of the most important topics in foods related to risk assessment, where the first chapter is an in depth coverage of predictive microbiology. This is followed by two chapters on the modeling of heat transfer, first on chilling and freezing followed by a chapter on heat sterilization. This is continued with a chapter on shelf–life prediction in chilled foods which reviews some of the practical modeling available to food processors for this important application. Two chapters cover the modeling of constituents in or from packaging materials starting with an overview of applications useable for the assessment of human exposure to contaminants from food contact materials, followed by a chapter on the modeling of migration of constituents from packaging materials.

The third section of the book is devoted to environmental issues and topics related to food raw materials as they may be affected by environmental pollutants or compounds like antibiotics. Most of these issues have originally been studied in other disciplines like pharmacy or veterinary science. However, the fate of antibiotics in meat products and pesticides and other pollutants in the environment are very important with respect to food safety and wholesomeness. Predictive modeling in these areas is therefore a very important issue for food scientists. The section is headed of with a chapter on the fate of antibiotics in animal tissue followed by three chapters where pollution is the central subject, one with a general view of the fate of pollutants, explaining the principles of transport and reactions underlying the fate of pollutants. The third chapter exemplifies the fate of a specific group of pollutants, pesticides, having a very high importance in food wholesomeness and consumer interest and awareness. The final chapter in the book deals with how nature can react to environmental pollution by phytoremidiation of metal contaminated soils for improving food safety.

The approach to the various topics in the book differs according to how long they have been under sturdy. In subjects with relatively recent interest like predictive microbiology and fate of pollutants the authors present an overview of current research and point out research needs. On the other hand the focus is on industrial application in subjects that have been under investigation for longer periods like heat treatment of foods. Many of the authors point out specific references for further study and all cite recent research in there approach.

March, 2008   Rui Costa and Kristberg Kristbergsson

Contents

Part III   Environment and Raw Food Production

Contributors

Katleen Baert
Ghent University, Department of Food Safety and Food Quality, Coupure Links 653, 9000 Gent, Belgium

Manuel Benlloch
Departamento de Agronomía, Escuela Técnica Superior de Ingenieros Agrónomos y Montes, Edif. C-4, Campus de Rabanales, Universidad de Córdoba, 14071 - Córdoba, Spain
ag1bemam@uco.es

Teresa R. S. Brandão
College of Biotechnology, Catholic University of Portugal, Rua Dr. António Bernardino de Almeida, 4200-072 Porto, Portugal
tsbrandao@mail.esb.ucp.pt

Johan Debevere
Ghent University, Department of Food Safety and Food Qualityity, Coupure Links 653, 9000 Gent, Belgium

Frank Devlieghere
Ghent University, Department of Food Safety and Food Quality, Coupure Links 653, 9000 Gent, Belgium
Frank.Devlieghere@UGent.be

Amílcar C. Falcão
Faculty of Pharmacy, University of Coimbra Rua do Norte, 3000-295 Coimbra, Portugal
acfalcao@ff.uc.pt

Maria de Fátima Poças
Packaging Department – College of Biotechnology, Catholic University of Portugal, Rua Dr. António Bernardino de Almeida, 4200-072 Porto, Portugal
mfpocas@esb.ucp.pt

Kjell François
Ghent University, Department of Food Safety and Food Quality, Coupure Links 653, 9000 Gent, Belgium

Vassilis Gekas
Department of Environmental Engineering, Technical University of Crete,
Polytechnioupolis, GR-73100, Chania-Crete, Greece
vgekas@enveng.tuc.gr

Gudmundur Gudmundsson
Lýsi hf, Fiskislóð 5-9, 101 Reykjavík, Iceland
gg@lysi.is

Christian James
Food Refrigeration and Process Engineering Research Centre, University of
Bristol, Churchill Building, Langford, Bristol, BS40 5DU, United Kingdom
chris.james@bristol.ac.uk

Stephen James
Food Refrigeration and Process Engineering Research Centre, University of
Bristol, Churchill Building, Langford, Bristol, BS40 5DU, United Kingdom

Veselka Kamburova
Rousse University, Studentska Str. 8, Rousse, 7017, Bulgaria

Laurence Ketteringham
Food Refrigeration and Process Engineering Research Centre, University
of Bristol, Churchill Building, Langford, Bristol, BS40 5DU, United Kingdom

Kristberg Kristbergsson
Department of Food Science, Faculty of Food Science and Nutrition, University
of Iceland, Saemundargata 6, 101 Reykjavík, Iceland
kk@hi.is

Bruno De Meulenaer
Ghent University, Department of Food Safety and Food Quality, Coupure Links 653,
9000 Gent, Belgium
Bruno.DeMeulenaer@UGent.be

R. Dios-Palomares
Departamento de Estadística, Escuela Técnica Superior de Ingenieros
Agrónomos y Montes, Edif. C-2, Campus de Rabanales, Universidad de Córdoba,
14071 - Córdoba, Spain
ma1dipar@uco.es

Silvia Palpacelli
Food Refrigeration and Process Engineering Research Centre, University
of Bristol, Churchill Building, Langford, Bristol, BS40 5DU, United Kingdom

Ioanna Paraskaki
Department of Environmental Engineering, Technical University of Crete,
Polytechnioupolis, GR-73100, Chania-Crete, Greece

Ioannis Th. Polyrakis
Department of Environmental Engineering, Technical University of Crete, Polytechnioupolis, GR-73100, Chania-Crete, Greece
IPOLIRAKIS@ate.gr

Enrique D. Sancho
Departamento de Microbiología, Escuela Técnica Superior de Ingenieros Agrónomos y Montes, Edif. C-6, Campus de Rabanales, Universidad de Córdoba, 14071-Córdoba, Spain
edsancho@uco.es

Stefan Shilev
Department of Microbiology and Environmental Biotechnologies, Agricultural University – Plovdiv, 12 Mendeleev Str., 4000-Plovdiv, Bulgaria
stefan.shilev@au-plovdiv.bg

Cristina L.M. Silva
College of Biotechnology, Catholic University of Portugal, Rua Dr. António Bernardino de Almeida, 4200-072 Porto, Portugal
clsilva@esb.ucp.pt

Ivanka Zheleva
Rousse University, Studentska Str. 8, Rousse, 7017, Bulgaria
vzh@abv.bg

Part I
Introduction

Chapter 1
Introduction to Integrated Predictive Modeling

Teresa R. S. Brandão and Cristina L.M. Silva

1.1 Introduction

When a mathematical model is properly developed, it is a potential tool for process design, assessment, and optimization. Using a mathematical expression that predicts a real observation with accuracy and precision is an efficient way to develop new products and to control systems. However, to attain a convenient model, a lot of well-guided experimental effort should be expended and the model should be validated. One should never forget that the model predicts one response in the range of experimental conditions tested and care should be taken when extrapolating to other operating conditions.

This chapter provides an introductory approach to concepts and methods involved in mathematical modeling, with particular focus on modeling quality and safety of food products.

1.2 Basic Concepts and Methods

Modeling is the use of mathematical language to describe the behavior of a system. A model is a mathematical expression that relates dependent variables(s) to independent variables(s). This relationship involves constants (i.e., parameters) that depend on intrinsic and/or extrinsic factors. In general, an observed response can be written as

$$y_i = f\left(x_{ij}, \theta_k\right) + \varepsilon_i,$$ (1.1)

where $i = 1, 2, \ldots, n$ is the number of experimental observations/runs, $j = 1, 2, \ldots, v$ is the number of independent variables, and $k = 1, 2, \ldots, p$ is the number of model

Cristina L.M. Silva (✉)
College of Biotechnology, Catholic University of Portugal, Rua Dr. António Bernardino de Almeida, 4200-072, Porto, Portugal

R. Costa, K. Kristbergsson (eds.), *Predictive Modeling and Risk Assessment*, DOI: 10.1007/978-1-387-68776-6, © Springer Science+Business Media, LLC 2009

parameters. y_i represents a measured response at the ith experimental observation (e.g., dependent variable), x_{ij} are fixed j independent variables that define the experimental conditions at observation i, θ_k are unknown parameters, and f is the mathematical form of the model considered. ε_i represents an independent experimental error (from a normal distributed error population with mean equal to zero and constant variance).

The mathematical expressions used in process modeling can be derived on the basis of fundamental reasoning or empirical description. The first ones are called mechanistic models, since they are based on knowledge of the fundamental mechanisms involved in a process. On the other hand, empirical modeling is a black-box approach, meaning that no concern is given to the theory underlying the phenomena, and the objective is merely a convenient prediction of the observation.

Particular types of models that can be considered neither mechanistic nor empirical are ones based on probability distributions, also referred to as stochastic models. In those models, one characteristic of the system is assumed to follow a preestablished probability density function (normal, Weibull, and logistic are examples of the most used lifetime distributions).

The mathematical models chosen should predict the response variable(s) accurately and the choice depends on the adequacy of the model to describe the process, and also on the quality of the parameters. Even the most complex model cannot yield good predictions if the model parameters are not estimated with accuracy and precision.

1.3   Experimental Design

In mathematical modeling, the very first step is the observation of a real occurrence and experimental data gathering. The information contained in data is established at the moment they are collected and, therefore, only an assertive selection of an experimental data pattern can provide good results. This is the objective of the *experimental design*.

The experimental designs resulting from the application of intuitive thinking to data collection are often called *heuristic designs* (i.e., based on "common sense"). Heuristic designs with sampling times that are equidistant (i.e., independent variable), are certainly the preferred ones. Even so, one should be provident in the time-extension of the whole experiment. This aspect is particularly important in kinetic studies where the results can be limited if one of the extreme situations occurs: experimental points cover only a small extension of the process, or there is a deficient amount of experimental data in early phases. In processes of great variation in the early stages before stabilization, data can be collected at equidistant points on the logarithmic time scale (Boulanger and Escobar, 1994). This obviously implies that more samples are collected at short times than at longer ones. If the experimenter has a preliminary idea about the process and the model to be used, the option can be a design with sampling times chosen so that the expected change in the response between two consecutive times is constant.

Owing to their characteristics and simplicity, heuristic designs are commonly and unquestionably used. Those designs have also the ability to test model adequacy in fitting experimental data.

If the experimenter is confident about the capability of a mathematical model in a process description, more complex experimental designs with challenging purposes may be developed. Experimental sampling and conditions may be planned to attain the most precise statistical inference, this being the concept of optimal design. Optimal designs rely on statistical theories and on different criteria. One of the most used criteria is the one that minimizes the confidence region of the parameters and, therefore, maximizes the parameters' precision, known as D-optimal design (see Steinberg and Hunter 1984 for an overview of optimality criteria). Box and Lucas (1959), in a pioneer work, illustrated clearly the applicability of D-optimal designs to simple first-order-decay kinetics. This work was the inspiration to design optimal experiments in different research fields. In food engineering, D-optimal experimental design was applied to kinetics of thermal degradation (Cunha et al., 1997; Cunha and Oliveira, 2000; Frías et al., 1998), Fickian mass-transfer kinetics (Azevedo et al., 1998; Brandão and Oliveira, 2001), systems described by the Weibull function (Boulanger and Escobar, 1994; Cunha et al., 2001), and enzyme-catalyzed reaction kinetics (Malcata, 1992).

The combination of a heuristic design and an optimal criterion with the objective of both testing model adequacy and attaining parameter precision is one interesting concept. This can be called a compromise design criterion (Brandão, 2004), and is certainly an attractive area to be exploited in the design of experiments.

Experiments may also be designed with the purpose of studying the effect of factors or treatments on an observation. The first statistician to consider such a method for the design of experiments was Ronald A. Fisher (1890–1962). Inspired by problems in agricultural experimentation, he realized that some important questions could not be answered because of the inherent weakness of experimental planning. Fisher was a pioneer in introducing the concepts of randomization and analysis of variance to the design of experiments. Experiments were designed in such a way that they differed from each other by one or several factors or treatments applied to them. The observed differences in the experimental outcomes could then be attributed to the different factors or combinations of factors, by means of statistical analysis. This was remarkable progress over the conventional approach of varying only one factor at a time in an experiment, which was a somewhat and in many situations a wasteful procedure. The classic book of Box et al. (1978) is a valuable compendium of experimental design procedures, with practical examples and clear explanations of how to use experiments to improve results.

Experimental design is definitely the most valuable aspect of statistical methodology. If sampling points and conditions are wisely selected, model adequacy can be tested, parameter estimation may be improved, and a rapid and clear evaluation of important variables affecting a system can be attained. However, no sophisticated and complex data analysis procedures will provide good results when experimental data are poor.

1.4   Data Analysis

After experimental data have been collected, data analysis is the following step. A mathematical skeleton based on theory, analogy, or inspection has to be chosen, and model parameters should be estimated by fitting the model to experimental points. Statisticians describe this topic as regression analysis, which follows identical steps independent of the model complexity assumed. The first step is the definition of a function, which should be a suitable measure of the deviation between the experimental data and the values given by the model, usually referred to as an objective function. The next step is the calculation of the parameters that minimize or maximize the selected objective function. This optimization procedure is the core of parameter estimation. Least-squares estimation, maximum likelihood estimation, and Bayesian estimation (Bard, 1974; Seber and Wild, 1989) are examples of well-known methods.

The least-squares method is undoubtedly the oldest and most used regression scheme. Contributing to its popularity is the simplicity of the mathematical principle, which basically consists in finding values for the parameters that minimize the sum of squares of deviations between the experimental values and the ones predicted by the mathematical model, avoiding the cognizance of the probability distributions of the observations.

The least-squares regression becomes complex if the model assumed is nonlinear (in the parameters). In such a situation the parameter estimation is not straightforward, since the minimization of the residuals, which is mathematically attained by calculation of the first and second derivatives of the model function with respect to the parameters, depends on the parameters themselves. In such circumstances, iterative methods for function minimization are required. Johnson and Frasier (1985) discussed the advantages and disadvantages of the numerical schemes commonly used (Nelder–Mead simplex concept, Newton–Raphson, Gauss–Newton, steepest descendent, and Marquardt–Levenberg).

If a model is adequate for fitting data, the residuals should reflect the properties assumed for the error term ε_i (Eq. 1.1). This is the basal concept underlying analysis of the residuals, which is a very simple and effective tool for detecting model deficiencies (Chatterjee and Price, 1991). The statistic runs test, which analyses the way experimental data are distributed along the model curve, can be used to validate the randomness of the residuals (Van Boekel, 1996). Randomness, however, is commonly assessed by plotting the residuals in a time sequence or against the estimated response of the model (Box et al., 1978). This plot also allows one to detect heteroscedasticity, i.e., inconstant variance of the errors. Unfortunately, many scientists forget examination of residuals and confine the evaluation of the regression quality to the correlation coefficient R (or the coefficient of determination, R^2), which is not the most effective evaluation in nonlinear regression analyses.

The prime objective of regression analysis is to provide estimates of the unknown model parameters. For a full parameter estimation, collinearity and precision of the estimates must be quantified. Collinearity (interdependency or correlation) of the parameter estimates may affect the optimization procedure inherent to the regression analysis and it can be evaluated by the correlation coefficient between the estimates (Bard, 1974). Collinearity can be reduced by model reparameterization, thus improving the estimation procedure. The Arrhenius equation expressed in terms of a finite reference temperature is the most well-known and used reparameterized model. When the reference temperature equals the average temperature of the temperature range considered, the best results are obtained.

Precision of the estimates can be evaluated by the confidence intervals (the intervals within which the true parameter values are expected to lie, for a chosen significance level). In multiparametric models, and owing to collinearity of the estimates, individual confidence intervals do not provide an adequate inference of the results. A better approach is the construction of a joint confidence region, where parameters probably exist together at a specific confidence level (Draper and Smith, 1981; Bates and Watts, 1988; Seber and Wild, 1989).

1.5 The Role of Modeling in Food Quality and Safety

In the last few years, increased attention has been given to food quality and safety, not only from a consumer's perspective but also by governmental entities. Consumers want food products to be safe and convenient to use, still having all the qualities of a fresh product. Foods often suffer during processing, which has three major aims: (1) to make food safe, (2) to provide products of the highest-quality attributes, and (3) to make food into forms that are more convenient or more appealing.

Safety assurance involves careful control of the whole process till the product reaches the consumer. Safety includes control of both chemical and microbiological characteristics of the product. From a chemical viewpoint, safety is generally related to keeping undesirable compounds such as pesticides, insecticides, and antibiotics out of the food supply chain. Most food processes deal with microbial control and often have as the objective the elimination or inactivation of microorganisms, or prevention of their growth. Processes that are aimed at prevention of growth include refrigeration, freezing, drying, and control of water activity, such as by addition of salts, sugars, polyols, etc. Concerning elimination of microorganisms, cleaning and sanitizing processes can be applied. With the aim of killing or inactivating microbes, thermal processes, such as pasteurization and sterilization, are often used.

Quality of a food product is a broad concept, which involves maintenance or improvement of key attributes, including, among others, nutrient retention, color, flavor, and texture. For maximum quality maintenance, it is essential to control microbial spoilage, enzymatic activity, and chemical degradation. Thermal processes, when conveniently applied, are efficient in microbial reduction and enzymatic inactivation, thus resulting in products with extended shelf life. Nevertheless, quality attributes may be negatively affected, especially owing to thermal impact at cellular tissue level.

More recently, alternative nonthermal technologies have been emerging in the food processing domain. Processes such as ultrasound, high pressure, and treatments with ozone and with ultraviolet radiation may be used to achieve efficient microbial reduction and improved quality retention of food products.

1.5.1 Food Safety

In the area of food safety, *modeling* has been gaining importance. The preconceived idea that microbiology is a science quite distant from mathematics or statistics concepts is becoming obsolete. Food scientists have been intensifying their dedication to the modeling of microbial growth and/or inactivation. This is certainly a cooperative effort between microbiologists and mathematicians and engineers. Pioneers designated this area as *"predictive microbiology,"* which emerged in the 1980s.

With the purpose of illustrating researchers' dedication to microbial kinetics modeling, published works containing "predictive microbiology" as a keyword were enumerated since the suggestion in 1986 of this designation by McMeekin and Olley (1986) (ISI Web of Knowledge[SM] 3.0, 2005). In the last 5 years the number of works doubled, showing the increased effort devoted to this field (see Fig. 1.1).

The groundwork of conjoining microbial behavior and mathematical transcription was probably laid by Esty and Meyer (1922). They were the first to describe

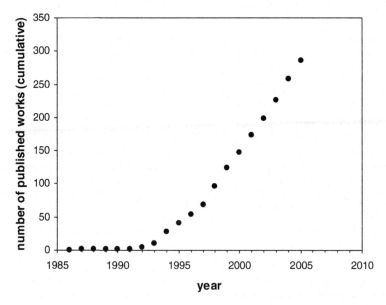

Fig. 1.1 Number of published works (cumulative) that contain "predictive microbiology" as a keyword, from 1986 till 2005 (ISI Web of Knowledge[SM] 3.0, 2005)

the thermal death of *Clostridium botulinum* spores in low-acid canned foods, by a simple log-linear relation with time (i.e., at a given temperature, the specific death rate of the bacteria is constant with time). This classical and dogmatic linear approach is an analogy of microbial inactivation kinetics to a first-order reaction, assuming constant environmental factors, such as temperature, pH, or water activity.

The influence of those factors on the inactivation rate constant (k) has been widely studied. When the temperature effect is to be assessed, the Arrhenius model is the one most commonly used. It was first derived on the basis of the application of thermodynamics to the analysis of the sensitivity of reaction rates to temperature (evaluated by an activation energy), and it has been applied empirically to the most diverse processes in numerous fields of investigation. Adaptations and extensions of the Arrhenius model to microbial death kinetic parameters, aiming at including other factors besides temperature, are discussed in other chapters. Other mathematical expressions that contain intrinsic and/or environmental effects on microbial kinetic parameters are also presented.

Classically, microbiologists prefer to quantify the inactivation behavior of microorganisms by using the decimal reduction time definition, D value (i.e., the time required for 1-log microbial reduction), in the first-order kinetics. It is widely accepted that the logarithm of the number of viable microorganisms decreases linearly with time, when unfavorably high temperatures are imposed. The slope of the line is the reciprocal of D, thus being an evaluation of the microbial heat resistance. The dependence of D on temperature results in the so-called Bigelow model, i.e.,

$$D = D_{ref}10^{(T_{ref}-T)/z},\tag{1.2}$$

where D_{ref} is the decimal reduction time at a reference temperature, T_{ref}, and z is the thermal death time parameter, defined as the temperature rise required for a tenfold decrease of D). This model, which is the basis for designing thermal processes in industrial canning of low-acid foods, was developed from experimental observations made by Bigelow (1921). Several works have suggested modifications to the Bigelow model that include pH and water activity effects on microbial heat resistance (Gaillard et al., 1998; Mafart and Leguérinel, 1998; Leguérinel et al., 2005).

The Bigelow and Arrhenius models are frequently involved in the never-ending discussion of which one is better than the other. One can say that more important than the approach selected is the quality of the parameter estimation, which is directly dependent on the experimental design applied and the data analysis chosen, as discussed in the previous section.

In cases where constant specific death rates are observed, k (from Arrhenius) and D (from Bigelow) are correlated by the expression $D = \ln 10/k$.

No extrapolations should be done when the logarithm of the microbial population size does not vary linearly with time.

In spite of the linear thermal death behavior having been accepted for almost one century, deviations from such tendency are often reported in the literature.

Generally, the curves may exhibit a delayed initial period prior to the exponential phase, often referred to as a *shoulder* (or *lag*; this designation is more commonly applied for bacterial growth) and/or a tailing phenomenon. The occurrence of a shoulder may be explained by nonhomogeneity of the heat sensitivity/resistance of cells, and can be related to the physiological state of the cells. In relation to the tail, researchers still do not agree whether the phenomenon is simply an experimental drawback in enumeration of cells (i.e., the tails occur at low cell concentration, often around the detection level threshold) or is actually a consequence of a heat-resistant residual population.

Modeling nonlinear inactivation kinetics is obviously a more complex task. The works of Xiong et al. (1999) and McKellar and Lu (2004) provide an overview of the models that have been used to describe sigmoid microbial inactivation. Models inspired and derived on the basis of Baranyi (Baranyi et al., 1993; Baranyi and Roberts, 1994), Gompertz (Gibson et al., 1987; Bhaduri et al., 1991; Linton et al., 1995), and logistic (Gibson et al., 1987; Zwietering et al., 1990) equations are the ones most often applied. However, most of the reported works did not use those equations in the most convenient form, neglecting the biological meaning of the parameters. Insightful information can be obtained with reparameterized forms. Estimation of the shoulder phase is frequently mistreated, or less reliable estimations are often used. There is still a lack of guided work to assess and include in predictive models factors that might affect this transition period.

Till the last decade, the greatest modeling effort was given to microbial inactivation under constant (or static) environmental conditions. From a realistic point of view this is somewhat restrictive, since the majority of thermal processes occur under time-varying environmental conditions, and kinetic parameters obtained under such circumstances may differ from the ones estimated under static conditions, which compromises safety control. Researchers are conscious of these difficulties and are trying to extend and structure models to deal with predictions in the time-variant domain (Van Impe et al., 1992; Geeraerd et al., 2000; Peleg et al., 2001; Bernaerts et al., 2004).

Temperature is doubtless the most studied factor that influences microbial survival. However, alternative nonthermal technologies have been emerging, a consequence of consumers' demand for fresh-like and minimally processed foods. High-pressure processes, ultraviolet light and ultrasound treatments, and treatments with ozone or pulsed electric fields are exciting areas in which predictive microbiology ought to be extended.

The book edited by McKeller and Lu (2004) is a valuable resource that covers the basics of modeling microbial responses in foods, with reflections and discussions on challenging aspects of predictive microbiology.

Predictions of microbial responses obtained from precise and accurate mathematical models that quantify the effect of preservation technologies on microbial populations can be helpful in the design, control, and optimization of industrial food processes and in the development of new processing operations. However, one should be aware of the groundwork required. Contributing to this are the complexity and variability of foods, the diversity of target food/microorganisms that might be involved, and the specific conditions required for each process, thus resulting in considerable experimental and time-consuming effort and financial requirements, not compatible with the priority of industrial management.

In addition to microbiological risks related to food-borne diseases, contamination of food by chemical compounds is a worldwide public health concern. Chemical risks for foods may be associated with food additives and contaminants that result from improper food manufacturing and processing. In some cases of foods processed at high temperatures, there is evidence that toxic compounds can be formed (e.g., in the case of acrylamide formation in the Maillard reaction; Mottram et al., 2002).

The migration of chemicals from food contact materials may also be the source of human health problems. Polymeric materials used in packages or package seals may be the origin of such contamination. An example is semicarbazide, a dangerous cancerous substance, whose presence in food products is linked to glass jars with metal lids that have foamed plastic seals (including those used for baby foods). The development of models for these sorts of processes to predict the concentration of these compounds in food is crucial for food safety as well. Mathematical fundamentals of the physical processes which govern the migration of components from packaging materials are discussed in another chapter.

The use of pesticides or other agrochemicals and animal drugs is a potential threat when not properly controlled. Food contamination may also occur through environmental pollution of air, water, and soil. Persistent organic pollutants, such as toxic metals, dioxins, and polychlorinated biphenyls, are dangerous chemicals to which the World Health Organization devotes particular attention. Environmental pollution from pesticides is discussed in another chapter, where the grounds for understanding the principles of the transport and reaction of pollutants, aiming at modeling the phenomena, are given.

1.5.2 Food Quality

The quality of foods depends implicitly on their attributes before processing and storage, and is affected by the extent and conditions of those operations. Processes are applied with the purpose of eliminating food-borne pathogens and inhibiting spoilage microorganisms or enzymes responsible for deterioration of quality, but, ironically, quality might be considerably affected by the process conditions themselves.

Predictions of food quality losses during processing and storage have been intensively investigated throughout the years. The concept of reaction rate has been widely and empirically used to describe changes in time (t) of the most diverse quality features (such as nutrient or sensory content, color, and texture), designated generally by Q in the following equation:

$$\frac{dQ}{dt} = -kQ^n, \tag{1.3}$$

where k is the rate constant and n the order of the reaction. Usually the kinetics of quality changes are assumed to be of zero or first order (Van Boekel, 1996).

Equation 1.3 can be used in the form of a reversible first-order model (Levenspiel, 1972):

$$\frac{d(Q-Q_{eq})}{dt} = -k(Q-Q_{eq}), \tag{1.4}$$

Q_{eq} being the equilibrium value. This model is also designated as the fractional conversion model, and has been used to model kinetics of color changes in thermally processed fruit and vegetable products (Steet and Tong, 1996; Ávila and Silva, 1999).

Quality can be affected by enzyme activity, responsible for the catalysis of a large number of reactions that cause off-flavors and off-colors, especially in raw and frozen vegetables that were subjected to a preblanching treatment (López et al., 1994). Hence, thermal inactivation of some sorts of enzymes may increase the products' shelf life and is frequently used as an index for the adequacy of the thermal process. Ling and Lund (1978) proposed a biphasic first-order model to describe isoenzyme thermal inactivation kinetics when different heat-resistance fractions are observed:

$$Q_E = Q_{E_{01}} e^{-k_1 t} + Q_{E_{02}} e^{-k_2 t}. \tag{1.5}$$

Q_E corresponds to the enzyme activity (the subscript 0 indicates the initial value). The indexes 1 and 2 are indicative of the heat-labile and the heat-resistant fractions, respectively. This model was used to describe mathematically the thermal inactivation of peroxidase in different vegetables, such as broccoli, green asparagus, and carrots (Morales-Blancas et al., 2002).

Besides the proven efficiency of thermal treatments in microbial and enzymatic inactivation, sensory food attributes are negatively impacted at the cellular tissue level. With the increased attention given to minimally processed foods, with maximum quality retention, researchers are studying nonthermal treatments, often complementary to classical technologies. With this purpose, Cruz et al. (2006) combined heat and ultrasound treatments, aiming at inactivating peroxidase in watercress, and used a biphasic model as shown in Eq. 1.5 to model the kinetics.

Stochastic approaches using Weibull distributions are also used to predict quality changes. For any quality attribute Q, the kinetics of degradation can be described by the function

$$\frac{Q}{Q_0} = 1 - e^{\left(-\frac{t}{\alpha}\right)^\beta}, \tag{1.6}$$

where α is a scale parameter that is associated with the process rate (i.e., the reciprocal of the rate constant) and β is a shape parameter that is a behavioral index. If different values are assigned to β, Eq. 1.6 gives different curves, allowing the description of a wide variety of kinetic behaviors. The mathematical simplicity and the versatility in dealing with processes with high variability make the Weibull probability distribution an attractive model for use in quality predictions.

The process conditions, such as temperature, pressure, water activity, pH, concentration of solutes, and air composition, affect the rate constants of the models. Usually the Arrhenius model and Arrhenius-inspired mathematical expressions are used to describe their temperature dependence. The influence of the other factors is further included in the Arrhenius-model parameters.

Polynomial expressions with mixed effects are also empirically applied to describe the process effects on rate constants.

The derivative form of Eqs. 1.3 and 1.4 (or modifications of any other models to include continuous quality variations throughout time) should be used to predict quality attributes on an industrial scale, where foods are typically processed under nonisothermal or nonstationary conditions. The history of the environmental conditions affects quality, and model predictions would be improved if such a dependency were accounted for. This was already mentioned in Sect. 1.5.1, when food safety was discussed. The time-derivative model forms allow also predictions of quality throughout food distribution chains, where fluctuations of temperature occur.

When a food product is processed, food components may initiate a complex chain of degrading/forming reactions, which can compromise quality. If reactants and products can be measured at the same time, multiresponse modeling techniques provide a better approach (Box and Draper, 1965; Stewart et al., 1992), since kinetic information of all components is considered. This results in insightful parameter estimation and, consequently, in more accurate model predictions (Van Boekel, 1996). This approach contributes also to the understanding and clarification of the reaction mechanisms involved.

1.6   Integrated Approach of Phenomena

Consumers desire safe food products that retain maximum quality, have an extended shelf life, and are convenient to prepare. Achievement of this goal requires global knowledge of all physical and chemical phenomena involved. From a modeler's viewpoint, the mathematical description of the phenomena should be integrated to ascertain quality and safety – *integrated predictive modeling*.

Most of the common processes for food safety and preservation (e.g., blanching, pasteurization, sterilization, canning, baking, or freezing) involve heat transfer. In many thermal treatments, however, heat and moisture exchanges occur simultaneously, such as in drying, smoking, baking, and cooking.

Modeling of heat in foods and awareness about including mass transfer, when important, is covered in a later chapter. The assignment is not simple, but it is captivating. It becomes even more interesting when linked to kinetics of food quality and safety. For instance, when a solid food suffers pasteurization with hot dry air, the water activity at the surface varies considerably. For accurate microbial content predictions, heat, mass, and microbial inactivation kinetics should be modeled jointly. Basically, the temperature at the food surface should be estimated accurately by heat- and mass-transfer models. The temperature and other process factors will be used to estimate the microbial kinetic parameters of the models that describe the behavior of microorganisms in food, thus allowing safety inference. These modeling trails are important for the complete control of such pasteurization processes, or can be helpful in developing new food pasteurization equipment. However, a great amount of experimental work is required, since reliable data on the relationship between bacterial death and the surface temperature of real foods are fundamental. Scientists participating in a European project, funded by the European Commission under the Fifth Framework Program Quality of Life and Management of Living Resources program have recently presented results on predicting the reduction in microbes on the surface of foods during surface pasteurization (the BUGDEATH project QLRT-2001-01415; coordination by Food Refrigeration and Process Engineering Research Centre, University of Bristol, UK). During the project, a piece of laboratory equipment to create controlled surface temperature profiles on a variety of solid foods was conceived. The equipment was used to establish microbial death kinetics for bacteria on real food surfaces. The models were included in user-friendly software that can be used to predict microbial reduction at food surfaces, when a thermal treatment is applied (Gil et al., 2006).

Another interesting subject in which integrated predictive models can be valuable for food quality prediction and simulation is food drying. Ramos et al. (2005) presented the mathematical basis and considerations for integrating heat, mass-transfer, and shrinkage phenomena in solar drying of grapes, with the purpose of improving the predictive ability of models. As the authors mentioned, if microstructural changes are additionally correlated to a quality attribute, or a final sensory property, a complete integrated predictive method will be achieved and will certainly contribute to relate drying conditions and food quality.

These final examples show the valuable role of mathematical modeling in developing tools that will certainly be helpful for food process design, control, and optimization. Only well-guided and interdisciplinary work between statisticians, engineers, food scientists, and microbiologists will lead to this being achievable.

References

Ávila, I.M.L.B. and Silva, C.L.M. (1999) "Modelling kinetics of thermal degradation of colour in peach puree", *Journal of Food Engineering*, 39(2), 161–166.

Azevedo, I.C.A., Oliveira, F.A.R. and Drumond, M.C. (1998) "A study on the accuracy and precision of external mass transfer and diffusion coefficients jointly estimated from pseudo-experimental simulated data", *Mathematics and Computers in Simulation*, 48(1), 11–22.

Baranyi, J. and Roberts, T.A. (1994) "A dynamic approach to predicting bacterial-growth in food", *International Journal of Food Microbiology*, 23(3–4), 277–294.

Baranyi, J., McClure, P.J., Sutherland, J.P. and Roberts, T.A. (1993) "Modelling bacterial-growth responses", *Journal of Industrial Microbiology*, 12(3–5), 190–194.

Bard, Y. (1974) "Nonlinear Parameter Estimation". Academic Press, New York, USA.

Bates, D.M. and Watts, D.G. (1988) "Nonlinear Regression Analysis and its Applications". Wiley, New York, USA.

Bernaerts, K., Dens, E., Vereecken, K., Geeraerd, A.H., Standaert, A.R., Devlieghere, F., Debevere, J. and Van Impe, J.F. (2004) "Concepts and tools for predictive modelling of microbial dynamics", *Journal of Food Protection*, 67(9), 2041–2052.

Bhaduri, S., Smith, P.W., Palumbo, S.A., Turmer-Jones, C.O., Smith, J.L., Marmer, B.S., Buchanan, R.L., Zaika, L.L. and Williams, A.C. (1991) "Thermal destruction of *L. monocytogenes* in liver sausage slurry", *Food Microbiology*, 8, 75–78.

Bigelow, W.D. (1921) "The logarithmic nature of thermal death time curves", *Journal of Infectious Diseases*, 29, 528–536.

Boulanger, M. and Escobar, L.A. (1994) "Experimental design for a class of accelerated degradation tests", *Technometrics*, 36(3), 260–272.

Box, G.E.P. and Draper, N.R. (1965) "The Baysean estimation of common parameters from several responses", *Biometrika*, 52(3,4), 355–365.

Box, G.E.P. and Lucas, H.L. (1959) "Design of experiments in non-linear situations", *Biometrika*, 46(1,2), 77–90.

Box, G.E.P., Hunter, W.G. and Hunter, J.S. (1978) "Statistics for Experiments. An Introduction to Design, Data Analysis and Model Building". Wiley, New York, USA.

Brandão, T.R.S. (2004) "Application of non-isothermal methods to the estimation of mass transfer parameters: analysis of the effect of experimental design and data analysis on the precision and accuracy of the estimates", Ph.D. thesis, Escola Superior de Biotecnologia, Universidade Católica Portuguesa, Porto, Portugal.

Brandão, T.R.S. and Oliveira, F.A.R. (2001) "Design of experiments for improving the precision in the estimation of diffusion parameters under isothermal and non-isothermal conditions", *International Journal of Food Science and Technology*, 36(3), 291–301.

Chatterjee, S. and Price, B. (1991) "Regression Analysis by Example". 2nd ed. Wiley, New York, USA.

Cruz, R.M.S., Vieira, M.C. and Silva, C.L.M. (2006) "Effect of heat and thermosonication treatments on peroxidase inactivation kinetics in watercress (*Nasturtium officinale*)", *Journal of Food Engineering*, 72(1), 8–15.

Cunha, L.M. and Oliveira, F.A.R. (2000) "Optimal experimental design for estimating the kinetic parameters of processes described by the first-order Arrhenius model under linearly increasing temperature profiles", *Journal of Food Engineering*, 46(1), 53–60.

Cunha, L.M., Oliveira, F.A.R., Brandão, T.R.S. and Oliveira, J.C. (1997) "Optimal experimental design for estimating the kinetic parameters of the Bigelow model", *Journal of Food Engineering*, 33(1,2), 111–128.

Cunha, L.M., Oliveira, F.A.R., Aboim, A.P., Frías, J.M. and Pinheiro-Torres, A. (2001) "Stochastic approach to the modelling of water losses during osmotic dehydration and improved parameter estimation", *International Journal of Food Science and Technology*, 36(3), 253–262.

Draper, N. and Smith, H. (1981) "Applied Regression Analysis". 2nd ed. Wiley, New York, USA.

Esty, J.R. and Meyer, K.F. (1922) "The heat resistance of the spore of *Bacillus botulinus* and allied anaerobes. XI", *Journal of Infectious Diseases*, 31, 650–663.

Frías, J.M., Oliveira, J.C., Cunha, L.M. and Oliveira, F.A.R. (1998) "Application of *D*-optimal design for determination of the influence of water content on the thermal degradation kinetics of ascorbic acid in dry environments", *Journal of Food Engineering*, 38(1), 69–85.

Gaillard, S., Leguérinel, I. and Mafart, P. (1998) "Model for combined effects of temperature, pH and water activity on thermal inactivation of *Bacillus cereus* spores", *Journal of Food Science*, 63, 887–889.

Geeraerd, A.H., Herremans, C.H. and Van Impe, J.F. (2000) "Structural model requirements to describe microbial inactivation during a mild heat treatment", *International Journal of Food Microbiology*, 59, 185–209.

Gibson, A.M., Bratchell, N. and Roberts, T.A. (1987) "The effect of sodium-chloride and temperature on the rate and extent of growth of *Clostridium-botulinun* type-A in pasteurized pork slurry", *Journal of Applied Bacteriology*, 62(6), 479–490.

Gil, M.M., Pereira, P.M., Brandão, T.R.S., Silva, C.L.M., Kondjoyan, A., Van Impe, J.F.M. and James, S. (2006). "Integrated approach on heat transfer and inactivation kinetics of microorganisms on the surface of foods during heat treatments – software development", *Journal of Food Engineering*, 76, 95–103.

Johnson, M.L. and Frasier, S.G. (1985) "Nonlinear least-squares analysis", *Methods in Enzymology*, 117, 301–342.

Leguérinel, I., Spegagne, I., Couvert, O., Gaillard, S. and Mafart, P. (2005) "Validation of an overall model describing the effect of three environmental factors on the apparent *D*-value of *Bacillus cereus* spores", *International Journal of Food Microbiology*, 100, 223–229.

Levenspiel, O. (1972) "Interpretation of batch reactor data". In Chemical Reaction Engineering. 2nd ed. Wiley, New York, USA.

Ling, A.C. and Lund, D.B. (1978) "Determining kinetic parameters for thermal inactivation of heat-resistant and heat-labile isozymes from thermal destruction curves", *Journal of Food Science*, 43, 1307–1310.

Linton, R.H., Carter, W.H., Pierson, M.D. and Hackney, C.R. (1995) "Use of a modified Gompertz equation to model nonlinear survival curves for *Listeria monocytogenes* Scott A", *Journal of Food Protection*, 58, 946–954.

López, P., Sala, F.J., Fuente, J.L., Condón, S., Raso, J. and Burgos, J. (1994) "Inactivation of peroxidase, lipoxygenase, and polyphenol oxidase by manothermosonication", *Journal of Agricultural and Food Chemistry*, 42, 252–256.

Mafart, P. and Leguérinel, I. (1998) "Modelling combined effect of temperature and pH on the heat resistance of spores by a non-linear Bigelow equation", *Journal of Food Science*, 63, 6–8.

Malcata, F.X. (1992) "Starting *D*-optimal designs for batch kinetic studies of enzyme-catalyzed reactions in the presence of enzyme deactivation", *Biometrics*, 48(3), 929–938.

McKellar, R.C. and Lu, X. (eds) (2004) "Modelling microbial responses in food". CRC Press, Boca Raton, USA.

McMeekin, T.A. and Olley, J. (1986) "Predictive microbiology", *Food Technology in Australia*, 38(8), 331–334.

Morales-Blancas, E.F., Chandia, V.E. and Cisneros-Zevallos, L. (2002) "Thermal inactivation kinetics of peroxidase and lipoxygenase from broccoli, green asparagus and carrots", *Journal of Food Science*, 67, 146–154.

Mottram, D.S., Wedzicha, B.L. and Dodson, A.T. (2002) "Acrylamide is formed in the Maillard reaction", *Nature*, 419(6906), 448–449.

Peleg, M., Penchina, C.M. and Cole, M.B. (2001) "Estimation of the survival curve of *Listeria monocytogenes* during non-isothermal heat treatments", *Food Research International*, 34(5), 383–388.

Ramos, I.N., Brandão, T.R.S. and Silva, C.L.M. (2005) "Integrated approach on solar drying, pilot convective drying and microstructural changes", *Journal of Food Engineering*, 67, 195–203.

Seber, G.A.F. and Wild, C.J. (1989) "Nonlinear Regression". Wiley, New York, USA.

Steet, J.A. and Tong, C.H. (1996) "Degradation kinetics of green colour and chlorophylls in peas by colorimetry and HPLC", *Journal of Food Science*, 61(5), 924–931.

Steinberg, D.M. and Hunter, W.G. (1984) "Experimental design: review and comment", *Technometrics*, 26(2), 71–97.

Stewart, W.E., Caracotsios, M. and Sorensen, J.P. (1992) "Parameter estimation from multiresponse data", *AIChE Journal*, 38(5), 641–650.

Van Boekel, M.A.J.S. (1996) "Statistical aspects of kinetic modelling for food science problems", *Journal of Food Science*, 61(3), 477–485, 489.

Van Impe, J.F., Nicolaï, B.M., Martens, T., De Baerdemaeker, J. and Vandewalle, J. (1992) "Dynamic mathematical model to predict microbial growth and inactivation during food processing", *Applied and Environmental Microbiology*, 58(9), 2901–2909.

Xiong, R., Xie, G., Edmondson, A.S., Linton, R.H. and Sheard, M.A. (1999) "Comparison of the Baranyi model with the Gompertz equation for modelling thermal inactivation of *Listeria monocytogenes* Scott A", *Food Microbiology*, 16, 269–279.

Zwietering, M.H., Jongenburger, I., Rombouts, F.M. and Van't Riet, K. (1990) "Modelling of the bacterial growth curve", *Applied and Environmental Microbiology*, 56(6), 1875–1881.

Chapter 2
Risk Assessment: A Quantitative Approach

K. Baert, K. Francois, B. De Meulenaer, and F. Devlieghere

2.1 Introduction

A risk can be defined as a function of the probability of an adverse health effect and the severity of that effect, consequential to a hazard in food (Codex Alimentarius, 1999). During a risk assessment, an estimate of the risk is obtained. The goal is to estimate the likelihood and the extent of adverse effects occurring to humans due to possible exposure(s) to hazards. Risk assessment is a scientifically based process consisting of the following steps: (1) hazard identification, (2) hazard characterization, (3) exposure assessment and (4) and risk characterization (Codex Alimentarius, 1999).

During the *hazard identification*, biological, chemical and physical agents that are capable of causing adverse health effects and which may be present in a particular food or group of foods are identified (Codex Alimentarius, 1999). In the second step, the *hazard characterization*, the nature of the adverse health effects associated with the hazards is evaluated in a qualitative and/or quantitative way (Codex Alimentarius, 1999); therefore, a dose–response assessment should be performed. The dose–response assessment is the determination of the relationship between the magnitude of exposure (dose) to a chemical, biological or physical agent and the severity and/or frequency of associated adverse health effects (response). The overall aim is to estimate the nature, severity and duration of the adverse effects resulting from ingestion of the agent in question (Benford, 2001). *Exposure assessment* is defined as the qualitative and/or quantitative evaluation of the likely intake of the hazard via food as well as exposure from other sources, if relevant (Codex Alimentarius, 1999). For food, the level ingested will be determined by the levels of the agent in the food and the amount consumed. The last step, *risk characterization*, integrates the information collected in the preceding three steps. It interprets the qualitative and quantitative information on the toxicological properties of a

F. Devlieghere (✉)
Department of Food Safety and Food Quality, Ghent University, Coupure Links, 653,
9000 Ghent, Belgium

R. Costa, K. Kristbergsson (eds.), *Predictive Modeling and Risk Assessment*,
DOI: 10.1007/978-1-387-68776-6, © Springer Science+Business Media, LLC 2009

chemical with the extent to which individuals (parts of the population, or the population at large) are exposed to it (Kroes et al., 2002). In other words, estimating how likely it is that harm will be done and how severe the effects will be. The outcome may be referred to as a risk estimate, or the probability of harm at given or expected exposure levels (Benford, 2001).

Quantitative risk assessment is characterized by assigning a numerical value to the risk, in contrast with qualitative risk analysis, which is typified by risk ranking or separation into descriptive categories of risk (Codex Alimentarius, 1999). During quantitative risk assessment, a model is used to calculate (estimate) the risk based on the exposure and the response to the dose. Besides the quantitative risk assessment model, the exposure and the dose response can also be described by a model. To calculate the exposure to a microbiological hazard for example, a model can be used that predicts the growth during storage. Several methods can be used to estimate the risk, namely (1) point estimates or deterministic modelling; (2) simple distributions and (3) probabilistic analysis (Kroes et al., 2002). In a deterministic framework, inputs to the exposure and effect prediction models are single values. In a probabilistic framework, inputs are treated as random variables coming from probability distributions. The outcome is a risk distribution (Verdonck et al., 2001). The method chosen will usually depend on a number of factors, including the degree of accuracy required and the availability of data (Parmar et al., 1997). No single method can meet all the choice criteria that refer to cost, accuracy, time frame, etc.; therefore, the methods have to be selected and combined on a case-by-case basis (Kroes et al., 2002).

2.2 Deterministic Risk Assessment

Deterministic modelling (point estimates) uses a single 'best guess' estimate of each variable within the model to determine the model's outcome(s) (Vose, 2000). This method is illustrated with the calculation of the exposure to a contaminant, for example the mycotoxin patulin. A fixed value for food consumption (such as the average or maximum consumption) is multiplied by a fixed value for the concentration of the contaminant in that particular food (often the average level or permitted level according to the legislation) to calculate the exposure from one source (food). The intakes from all sources are then summed to estimate the total exposure.

Point estimates are commonly used as a first step in exposure assessments because they are relatively quick, simple and inexpensive to carry out. Inherent in the point-estimate models are the assumptions that all individuals consume the specified food(s) at the same level, that the hazard (biological, chemical or physical) is always present in the food(s) and that it is always present at an average or high concentration. As a consequence, this approach does not provide insight into the range of possible exposures that may occur within a population or the main factors influencing the results of the assessment. It provides limited information for risk managers and the public. The use of this method also tends to significantly overestimate or underestimate the actual exposure (Finley and Paustenbach, 1994). Using high-level values to represent either the food consumption or the hazard level may lead to high and often implausible overestimates of the intake. Point estimates are generally considered to be most appropriate for screening purposes (Parmar et al., 1997). If they demonstrate that the intake is very low in relation to the accepted safe level for the hazard or below a general threshold value, even when assuming high concentrations in the food and high consumption of this food, it may be sufficient to decide that no further exposure assessments are required (Kroes et al., 2000). To refine estimates of exposure, more sophisticated methods for integrating the food consumption and hazard level and more detailed data from industry, monitoring programmes, etc. are needed (Kroes et al., 2002).

2.3   Probabilistic Risk Assessment

In probabilistic analysis the variables are described in terms of distributions instead of point estimates. In this way all possible values for each variable are taken into account and each possible outcome is weighted by the probability of its occurrence.

Probabilistic methods may be used under a classical or a Bayesian statistical paradigm (IEH, 2000). This text will only discuss the classical view (Monte Carlo simulation).

Monte Carlo simulation has been extensively described in the literature (Cullen and Frey, 1999; Vose, 1996). The principle is illustrated in Fig. 2.1. One random sample from each input distribution is selected, and the set of samples is entered into the deterministic model. The model is then solved, as it would be for any deter- ministic analysis and the result is stored. This process or iteration is repeated

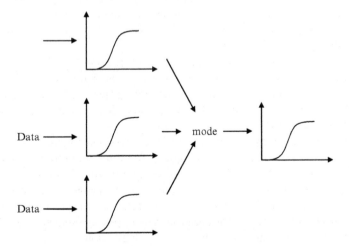

Fig. 2.1 The principle of a first-order Monte Carlo simulation (based on Verdonck, 2003)

several times until the specified number of iterations is completed. Instead of obtaining a discrete number for model outputs (as in a deterministic simulation), a set of output samples is obtained (Rousseau et al., 2001; Cullen and Frey, 1999). This method is described as a first-order Monte Carlo simulation or a one-dimensional Monte Carlo simulation.

2.3.1 Classical and Bayesian Probabilistic Risk Assessment

In 'classical' (also called 'frequentist') probabilistic risk assessment, probability is regarded as the frequency, that is how likely it is, based on repeated sampling, that an event will occur. Monte Carlo simulation is the method most commonly used for classical probabilistic risk assessment (IEH, 2000). It is a computational method developed over 50 years ago for military purposes, and which is used to achieve multiple samples of the input distributions, analogous to doing repeated experiments (IEH, 2000; Vose, 1996). Recently, attention has been paid to the Bayesian probabilistic approach. Bayesian statistical methods are based on Bayes's theorem, which was originally stated in 1763 by Thomas Bayes and was one of the first expressions, in a precise and quantitative form, of one of the modes of inductive inference. According to the Bayesian view, probability is regarded as a 'degree of belief', or how likely it is thought to be, based on judgement as well as such observational data as are available, that an event will occur. Bayesian theory requires a 'prior distribution' which is a probability density function relating to prior knowledge about the event at issue. This is combined with the distribution of a new set of experimental data in a likelihood framework (analogous to a probability distribution) to obtain the so-called posterior distribution of probability, the updated knowledge. Obtaining the posterior distribution can involve solving a complex integral, which is increasingly dealt with by means of Markov chain Monte Carlo methods. Thus, Monte Carlo methods are used both in Bayesian and classical probabilistic methods, although Bayesian analysis extends the use beyond simply simulating predictive distributions (IEH, 2000).

The process of setting up and running the models requires appropriate modelling software and a high level of computer processing power. There are a variety of risk analysis software products on the market, and these include software for modelling and distribution fitting. Examples of software products are @RISK, Crystal Ball and Fare Microbial (free software).

An important advantage of probabilistic risk assessments is that it permits one to consider the whole distribution of exposure, from minimum to maximum, with all modes and percentiles. In this way more meaningful information is provided to risk managers and the public. A second important advantage is the possibility to carry out a sensitivity analysis (see Sect. 2.5) (Finley and Paustenbach, 1994). An important disadvantage of the current probabilistic risk assessment procedures is the need for accurate prediction of the tails in a distribution (e.g. the fifth and 95th percentiles). These tails are very important when performing a probabilistic risk assessment because the largest-exposure concentrations (e.g. 95th percentile) will have first effects on the most sensitive population (e.g. fifth percentile) (Verdonck et al., 2001). Also the high degree of complication and time consumption are a disadvantage (Finley and Paustenbach, 1994). While probabilistic modelling confers many advantages, Burmaster and Anderson (1994) point out that the old computer maxim of 'garbage in, garbage out' (GIGO) also applies to this technique and that GIGO must not become 'garbage in, gospel out'. The reliability of the results of a probabilistic analysis is dependent on the validity of the model, the software used and the quality of the model inputs. The quality of the model inputs reflects both the quality of the data on which the input will be based and the selection of the distribution to represent the data in the model.

2.3.2   Quality of Data

Different sources of dietary information exist. In this context two methods, namely food supply data and individual dietary surveys, will be discussed. Food supply data are calculated in food balance sheets, which are accounts, on a national level, of annual production of food, changes in stocks, imports and exports, and agricultural and industrial use. The result is an estimate of the average value per head of the population, irrespective of, for instance, age or gender. Food supply data refer to food availability, which gives only a crude (overestimated) impression of potential average consumption. When these data are used in a risk assessment, there is high chance that the risk is overestimated (Kroes et al., 2002).

In contrast to food balance sheets, data from individual surveys provide information on average food and nutrient intake and their distribution over various well-defined groups of individuals. These data reflect more closely actual consumption (Kroes et al., 2002). However, it is important to note that individual dietary surveys will be associated with some degree of underreporting, since being engaged in a dietary survey affects customary consumption. This may potentially lead to some degree of underestimation of exposure and risk.

The distributions that are used as a model input can be a distribution function (parametric approach) or the data as such (non-parametric approach). For the parametric approach, the data are fitted to a distribution function, such as the normal distribution, the gamma distribution, the binomial distribution, etc. The distribution function that gives the best fit is used as a model input. For the non-parametric approach, the original data are used as an input for the model.

In the context of exposure assessments, 'simple distributions' is a term used to describe a method that is a combination of the deterministic and probabilistic approach. It employs a distribution for the food consumption, but it uses a fixed value for the level of the hazard. The results are more informative than those of the point estimates, because they take account of the variability that exists in food consumption patterns. Nonetheless, they usually retain several conservative assumptions (e.g. all soft drinks that an individual consumes contain a particular sweetener at the maximum permitted level; 100% of a crop has been treated with a particular pesticide; etc.) and therefore usually can only be considered to give an upper-bound estimate of exposure (Kroes et al., 2002).

2.4 Variability Versus Uncertainty

In most current probabilistic risk assessments, variability and uncertainty are not treated separately, although they are two different concepts. Variability represents inherent heterogeneity or diversity in a well-characterized population. Fundamentally a property of nature, variability is usually not reducible through further measurement or study (e.g. biological variability of species sensitivity or the variability of the amount of food consumed). As explained in Sects. 2.2 and 2.3, deterministic risk assessment does not take the variability into account, while probabilistic risk assessment includes variability in the calculations. In Fig. 2.2, the cumulative variability distributions are visualized as a full line (Verdonck et al., 2001).

Uncertainty represents partial ignorance or lack of perfect information about poorly characterized phenomena or models (e.g. measurement error), and can partly be reduced through further research (Cullen and Frey, 1999). In Fig. 2.2 the uncertainty is visualized as a band around the cumulative variability distribution function (dashed lines). For each percentile of the variability distribution, an uncertainty or confidence interval can be calculated (i.e. the uncertainty distribution) (Verdonck et al., 2001). For example, techniques such as bootstrapping can be used to get a measure of the uncertainty associated with the parameters of a distribution due to sample size (Cullen and Frey, 1999). The more limited the data set, the wider the confidence intervals will be.

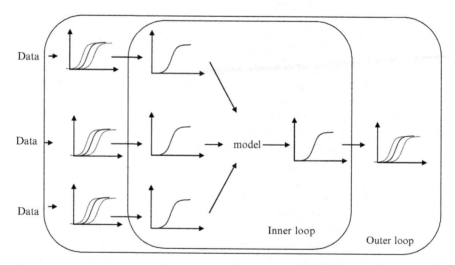

Fig. 2.2 The principle of a second-order Monte Carlo simulation (based on Verdonck, 2003)

To separate variability and uncertainty in a probabilistic risk assessment, a second-order or two-dimensional Monte Carlo simulation is developed (Burmaster and Wilson, 1996; Cullen and Frey, 1999). It simply consists of two Monte Carlo loops, one nested inside the other (Fig. 2.2). The inner loop deals with the variability of the input variables, while the outer one deals with uncertainty (Verdonck, 2003). First a distribution is selected randomly from every uncertainty band. Then this distribution is used in the inner loop and a first-order Monte Carlo simulation is carried out. The result is a distribution for the output of the model. This process is repeated several times. In this way a high number of output distributions is obtained and an uncertainty band can be constructed (outer loop).

2.5   Sensitivity Analysis

A sensitivity analysis is performed to determine which variables in the model have the biggest influence on the results. This is achieved by changing the value for each input variable in turn by a fixed amount (e.g. 10%) while keeping the others constant (IEH, 2000).

The results of sensitivity analysis permit risk managers to consider the relative merits of different strategies for reducing exposure in cases where levels of exposure are deemed unacceptably high. Because the probabilistic analysis provides information on the full distribution of exposures, the exposure assessor can determine how different scenarios will affect different sections of the distribution. For example, different scenarios of modelling nutrient supplementation or fortification can be applied to evaluate the impact at the lower tail of the distribution (possible inadequate nutrient intake) and the upper tail of the distribution (possible nutrient toxicity) (Kroes et al., 2002).

References

Benford D., Principles of risk assessment of food and drinking water related to human health. ILSI Europe Concise Monograph Series, 2001.

Burmaster D.E., Anderson P.D., Principles of good practice for the use of Monte Carlo techniques in human health and ecological risk assessment, *Risk Anal*, 14, 477–481, 1994.

Burmaster D.E., Wilson A.M., An introduction to second-order random variables in human health risk assessments, *Hum Ecol Risk Assess*, 2, 892–919, 1996.

Codex Alimentarius, Principles and guidelines for the conduct of microbiological risk assessment, CAC/GL-30, 1999.

Cullen A.C., Frey H.C., Probabilistic techniques in exposure assessment. A handbook for dealing with variability and uncertainty in models and inputs, Plenum, New York, 1999

Finley B., Paustenbach D., The benefits of probabilistic exposure assessment: three case studies involving contaminated air, water and soil, *Risk Anal*, 14, 53–73, 1994.

IEH (The Institute of Environment and Health), Probabilistic approaches to food risk assessment, 2000.

Kroes R., Galli C., Munro I., Schilter B., Tran L., Walker R., Wurtzen G., Threshold of toxicological concern for chemical substances present in the diet: a practical tool for assessing the need for toxicity testing, *Food Chem Toxicol*, 38, 255–312, 2000.

Kroes R., Müller D., Lambe J., Löwik M.R.H., van Klaveren J., Kleiner J., Massey R., Mayer S., Urieta I., Verger P., Visconti A., Assessment of intake from the diet, *Food Chem Toxicol*, 40, 327–385, 2002.

Parmar B., Miller P.F., Burt R., Stepwise approaches for estimating the intakes of chemicals in food, *Regul Toxicol Pharmacol*, 26, 44–51, 1997.

Rousseau D., Verdonck F.A.M., Moerman O., Carrette R., Thoeye C., Meirlaen J., Vanrolleghem P.A., Development of a risk assessment based technique for design/retrofitting of WWTPs, *Water Sci Technol*, 43, 287–294, 2001.

Verdonck F., Geo-referenced probabilistic ecological risk assessment, Ph.D. thesis, Ghent University, 2003.

Verdonck F., Janssen C., Thas O., Jaworska J., Vanrolleghem P.A., Probabilistic environmental risk assessment, *Med Fac Landbouw Univ Gent*, 66, 13–19, 2001.

Vose D., Quantitative risk analysis. A guide to Monte Carlo simulation modelling, Wiley, New York, 1996.

Vose D., Risk analysis: a quantitative guide, Wiley, New York, 2000.

Part II
Processing, Distribution and Consumption

Chapter 3
Predictive Microbiology

F. Devlieghere, K. Francois, A. Vermeulen, and J. Debevere

3.1 Introduction

Recent crises in the food industry have increased the awareness of the public of the food they eat. In the last few decades, dioxins and polychlorinated biphenyls (PCBs), but also microbial hazards such as *Listeria monocytogenes* or *Bacillus cereus* have reached the news headlines. The consumer has become more critical towards foods, demanding fresher, healthy, safe, and nutrition-rich food products.

One way to prove the microbial safety of a food product is by using laborious and time-consuming challenge tests. In these tests the shelf life of a specific food product can be assessed regarding spoilage and pathogenic microorganisms for a specific set of storage conditions. These methods were criticized for their expensive, time-consuming and noncumulative character (McDonald and Sun, 1999). As most food companies have an increasing number of different products, and storage conditions are different at each stage in the food chain, it is almost impossible to cover all these product/condition combinations using the classic challenge tests.

Predictive microbiology could be an answer to these problems. Once a model has been developed it will be the fastest way to estimate the shelf life according to microbial spoilage and to estimate the microbial safety of a food product. And within no time, this information can be provided for some different storage conditions or for some slight or more profound recipe changes. Moreover, in the last decade, computing power increased exponentially, while the price of a powerful machine was lowered, making it possible for everyone to access the current modeling methods.

Predictive food microbiology can be defined as the use of mathematical expressions to describe microbial behavior in a food product (Whiting and Buchanan, 1994). Predictive microbiology is based on the premise that the responses of populations of microorganisms to environmental factors are reproducible and that, by

F. Devlieghere (✉)
Department of Food Safety and Food Quality, Ghent University, Coupure Links 653, 9000 Ghent, Belgium

R. Costa, K. Kristbergsson (eds.), *Predictive Modeling and Risk Assessment,*
DOI: 10.1007/978-1-387-68776-6, © Springer Science + Business Media, LLC 2009

characterizing environments in terms of those factors that most affect microbial growth and survival, it is possible from past observations to predict the responses of those microorganisms in other, similar environments (Ross et al., 2000). This knowledge can be described and summarized in mathematical models which can be used to predict quantitatively the behavior of microbial populations in foods, e.g., growth, death, and toxin production, from knowledge of the environmental properties of the food over time.

Classically, the models can be divided into three groups, the primary models depicting the change in the number of bacteria over time, under a given set of conditions. When these conditions are favorable for the bacteria, the primary model will be a growth model, while under stressful conditions, the primary model will be an inactivation model. The secondary models deal with the effect of environmental conditions such as temperature, pH, a_w, or gas conditions on the primary model. In a tertiary model, the previous two models were brought together and implemented in a user-friendly software package.

In this chapter, we wish to provide an outline of the current modeling tools: the different types of primary and secondary models will be discussed, as well as some tertiary models, the model development procedures will be highlighted, the current deficiencies in predictive modeling will be revealed, and some illustrating examples will be presented.

3.2   Modeling the Growth of Microorganisms

Several different classification schemes are possible to group the models in the area of predictive modeling. Most often the system of Whiting and Buchanan (1993) is used, dividing the models into three groups, primary, secondary, and tertiary models, although some other segmentation criteria are also common. A distinction can be made between kinetic and probability models, between empirical and mechanistic models, and between static and dynamic models. In this section, these different groups will be discussed thoroughly and some examples will be given.

3.2.1   *Kinetic Versus Probability Models*

The first way to split up model categories within the area of predictive modeling is the difference between kinetically based models and probability based models. The choice of approach is largely determined by the type of microorganisms expected to be encountered and the number of variables. Kinetic models can predict the extent and rate of growth of a microorganism, while probability based models consider the probability of some event occurring within a nominated period of time. They are often used to model spore-forming bacteria, such as the probability of survival and germinating of spores in a canned product. Probability models can also be used to estimate the likelihood that a pathogen can be present at dangerous levels in a food product given a certain set of environmental conditions, or the likelihood that a pathogen is able or not able to grow in a certain environment.

Kinetic modeling focuses on the bacterial cell concentration as a function of time. Normally this can be depicted by a sigmoidal growth curve (Fig. 3.1), which is characterized by four main parameters: (1) the lag phase λ (h), which is the time that a microorganism needs to adapt to its new environment; (2) the growth speed μ (h^{-1}), which is correlated to the slope of the log-linear part of the growth curve; (3) the initial cell concentration N_0 (cfu/ml), and (4) the maximal cell concentration N_{max} (cfu/ml)

The effects of different environmental factors such as temperature, pH, and a_w on the parameter values of those sigmoidal curves are then modeled. Evaluation of this fitted sigmoidal curve may allow researchers to make predictions about the micro-organisms studied in a particular food system. In both modeling approaches, models are constructed by carefully evaluating a lot of data collected on increases in microbial biomass and numbers, under a studied criterion of intrinsic and extrinsic parameters. This allows researchers to make predictions about the lag phase, generation time, or exponential growth rate and maximum cell density of the microorganisms studied. Traditionally, the environmental variables studied are temperature, pH, and a_w, but also other important conditions such as the composition of the gas atmosphere, the type and concentration of acid, the relative humidity, the concentration of preservatives and/or antimicrobials, and the redox potential can affect the microbial growth and thus those factors can be included in a model.

Kinetic models are useful as they can be used to predict changes in microbial cell density as a function of time, even if a controlling variable, which can affect growth, is changing. However, the main drawback is that kinetic models are difficult to develop as they require a lot of data to be collected to model the interaction effects between the different environmental factors (McDonald and Sun, 1999). The main types of kinetic models will be discussed in Sect. 3.2.3.

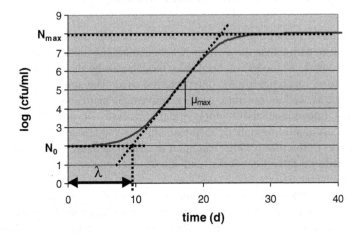

Fig. 3.1 Sigmoidal growth curve

Probability models are rather used for spore-forming bacteria and pathogens, as the focus here is on the probability that a certain phenomenon occurs. Originally, these models were concerned with predicting the likelihood that organisms would grow and produce toxins (e.g., for *Clostridium botulinum*) within a given period of time. Probability models can be a helpful tool for a food manufacturer to make an informed decision about product formulation, processing, packaging or storage of the products. They give appropriate information with regard to toxin production or spore germination in a food, but they provide little information on growth rate (Gibson and Hocking, 1997).

More recently, probability models have been extended to define the absolute limits for growth of microorganisms in specified environments, e.g., in the presence of a number of stresses which individually would not be growth limiting but collectively prevent growth (Ross et al., 2000). These models, also called "growth/no growth interface models", are situated on the transition between the growth models and the inactivation models. They will be further discussed in Sect. 3.3.3.

Ross and McMeekin (1994) suggested that the traditional division of predictive microbiology into kinetic models and probability models is artificial. They argued that the two types of modeling approach represent the opposite ends of a spectrum of modeling requirements, with research at both ends eventually coming together. This can be proved by the growth/no growth models, which are a logical continuation of the kinetic growth models. When environmental conditions become more stressful, growth speed slows down and the lag phase becomes longer until the growth speed drops to zero, or the lag phase tends to eternity. Within that twilight zone, the modeling needs shift from a kinetic model when the environmental conditions still support growth to a probability model when the growth/no growth interface is reached, demonstrating the near link between the two modeling types.

3.2.2 Empirical Versus Mechanistic Models

The difference between empirical models and mechanistic models might be even more difficult to define than the difference between kinetic models and probability models. An empirical model, or black box model, can be described as a model where the fitting capacities are the only criterion used. The aim is to describe the observed data as well as possible, using a convenient mathematical relationship, without any knowledge about biochemical processes or underlying cellular processes. Polynomial models or response surface models are the best-known representatives of this group. Other examples of empirical models are the Gompertz equation, artificial neural networks, and square-root models and their modified versions (see later).

Mechanistic models are situated at the other side of the spectrum. Here, the model is based on the understanding of the underlying biochemical processes and intercellular and intracellular processes that are controlling the cellular behavior. Hence, it can be understood that the actual microbial knowledge is insufficient to generate a complete mechanistic model. Therefore, very often, a model will be a combination of some mechanistic and some empirical components. Most researchers agree that (semi-)mechanistic models are inherently superior to empirical models as they give a better understanding of the cellular behavior, so these models are to be preferred as long as the fitting capacities are not endangered.

3.2.3   Primary, Secondary, and Tertiary Models

The most used and maybe the most evident classification within the predictive modeling area is the distinction between primary, secondary, and tertiary models, firstly presented by Whiting and Buchanan (1993).

3.2.3.1   Primary Models

Primary models describe the microbial evolution as a function of time for a defined set of environmental and cultural conditions. This section will focus on the primary growth models, but the same structure can be followed for inactivation modeling.

Classically, a sigmoidal growth curve is used to model the cell growth as a function of time, although a three-phase linear model is sometimes used as a good simplification (Buchanan et al., 1997a). In Fig. 3.2, a standard growth curve is depicted, plotted over the original dataset. The symbols show the data collected, while the full line represents the sigmoidal growth model. Mostly, growth curves are figured on a $\log_{10}$-based scale for the cell density as a function of time, as microbial growth is an exponential phenomenon, but sometimes a natural logarithm base is preferred.

Traditionally, the sigmoidal growth curve is described using four parameters: the exponential growth rate μ_{max}, the lag phase λ, the inoculum level N_0, and the maximal cell density N_{max}. The exponential growth rate is defined as the steepest tangent to the exponential phase, so is the tangent at the inflexion point, while the lag phase

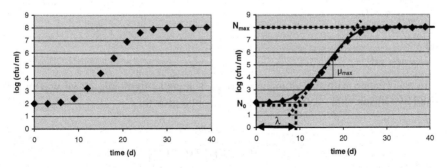

Fig. 3.2 Sigmoidal growth curve fitted to experimental data set

is defined as the time at which that extrapolated tangent line crosses the inoculum level (McMeekin et al., 1993). The generation time, the time necessary for a cell to multiply in hours or days, can be calculated from μ_{max}:

$$GT = \frac{\log_{10} 2}{\mu_{max}}, \tag{3.1}$$

where GT is the generation time.

During the short history of predictive modeling, several primary models were developed. The first growth model was described by Monod, and was based on a pure empirical observation that microbial cell growth is an exponential system.

$$N = N_0 e^{kt}. \tag{3.2}$$

This rather rudimentary model, where N is the cell density (cfu/ml), N_0 is the initial cell level (cfu/ml), $k = \ln 2/GT$ (h^{-1}), and t is time (h), is limited as only one growth parameter can be revealed: the growth rate μ_{max}, while no lag phase can be modeled. Also the maximal cell density is not taken into account, although cells cannot grow infinitely.

The nonlinear models were proposed in the 1980s. Gibson et al. (1987) introduced the Gompertz function to food microbiology, making it possible to express log(cfu/ml) as a function of time using the sigmoidal shape:

$$\log N(t) = A + D \exp\left\{\exp\left[-B(t-M)\right]\right\}, \tag{3.3}$$

where $N(t)$ is the cell density at time t, A is the value of the lower asymptote (N_0), D is the difference in value between the upper and the lower asymptotes ($N_{max}-N_0$), M is the time at which the growth rate is maximal, and B is related to the growth rate.

The parameters from the Gompertz model can be reparameterized to the classic primary growth curve characteristics:

$N_0 = A,$
$N_{max} = A + D,$
$\mu_{max} = BD/\exp(1),$
and
$\lambda = M - 1/B.$

In accordance with the Gompertz model, a logistic sigmoidal curve was proposed by Gibson et al. (1987) to predict microbial growth:

$$\log N(t) = \frac{D}{A + \exp\left[-B(t-M)\right]}. \tag{3.4}$$

The fitting results are similar to those of the Gompertz function, but the logistic model is a symmetrical model, while most growth curves are not; therefore the Gompertz model was preferred over the logistic model.

In the early 1990s, the focal point moved from the static primary models to the dynamic primary models. Van Impe et al. (1992) suggested a dynamic first-order differential equation to predict both microbial growth and inactivation, with respect to both time and temperature. This was one of the first models developed that was able to deal with time-varying temperatures over the whole temperature range of growth and inactivation. Also the previous history of the product can be taken into account. In the special case of constant temperature, the model behaved exactly like the corresponding Gompertz model.

A second dynamic growth model was developed by Baranyi and Roberts (1994). The model is based on a first-order ordinary differential equation:

$$\frac{\mathrm{d}x}{\mathrm{d}t} = \mu(x)x(t), \tag{3.5}$$

where $x(t)$ is the cell concentration at time t and $\mu(x)$ is the specific growth rate. If $\mu(x) = \mu_{max}$ = constant, then the equation describes the pure exponential growth.

This first-order differential equation was extended with two adjustment functions: an adaptation function $\alpha(t)$ describing the smooth transition from the inoculum level to the exponential growth phase and an inhibition function $u(x)$ describing the transition from the exponential growth phase to the stationary phase. So the structure for the growth model of Baranyi and Roberts is:

$$\frac{\mathrm{d}x}{\mathrm{d}t} = \mu_{max}\alpha(t)u(x)x(t). \tag{3.6}$$

The Model of Baranyi and Roberts

The basic model of Baranyi and Roberts is given by Eq. 3.6, but to apply the model, the adjustments functions $\alpha(t)$ and $u(x)$ have to be defined. The adaptation function is based on an additional parameter $q(t)$, representing the physiological state of the cells introduced. It creates an adaptation function that describes a capacity-type quantity expressing the proportion of the potential specific growth rate (which is totally determined by the actual environment) that is utilized by the cells. The process of adjustment (which is the lag period) is characterized by a gradual increase of $\alpha(t)$ from a low value towards 1:

$$\alpha(t) = \frac{q(t)}{1+q(t)}. \tag{3.7}$$

The inhibition function $u(x)$ can be based on nutrient depletion, which results in a Monod model (Eq. 3.8):

$$u(x) = \frac{S}{K_S + S}.$$

$$(3.8)$$

But nutrient depletion is only limiting at high cell concentrations and sometimes the growth inhibition starts much earlier. Therefore, a simple inhibition function was created, based on a maximum cell density parameter x_{max} and a curvature parameter m, characterizing the transition of the growth curve to the stationary phase (Eq. 3.9):

$$u(x) \approx 1 - \left(\frac{x(t)}{x_{max}} \right)^m.$$

$$(3.9)$$

For a specific set of conditions, the differential form of the Baranyi model can be solved, resulting in an explicit, deterministic, and static model, reparameterized to the classic growth curve parameters (Poschet and Van Impe, 1999):

$$N(t) = N_0 + \mu_{max} A(t) - \frac{1}{m} \ln \left(1 + \frac{e^{m\mu_{max}A(t)-1}}{e^{mN_{max}-N_0}} \right),$$

$$A(t) = t + \frac{1}{\mu_{max}} \ln \left(\frac{e^{-\mu_{max}t} + \left(e^{\lambda\mu_{max}-1} \right)^{-1}}{1 + \left(e^{\lambda\mu_{max}-1} \right)^{-1}} \right).$$

$$(3.10)$$

Nowadays, the growth model of Baranyi and Roberts is still one of the most used primary models. Many researchers have used the Baranyi model in specific microbial modeling applications and have found in comparison with the Gompertz function, and other models, that it gives satisfactory results (McDonald and Sun, 1999).

This success can be partly described to the program Microfit, where researches and producers can fit the Baranyi model easily to their datasets. The program is freeware that can be downloaded from the Internet. More information and the program can be found at http://www.ifrn.bbsrc.ac.uk/microfit.

3.2.3.2 Secondary Models

Secondary models model the effect of intrinsic (pH, a_w, etc.) and extrinsic (temperature, atmosphere composition, etc.) factors on the growth of a microorganism within a food matrix. Several modeling approaches have been designed, ranging from complete empirical models to more mechanistic oriented attempts with a lot of variations in model complexity. Four groups will be highlighted here: (1) the Arrhenius model and its modifications, (2) the Belehradek models, or square-root models, (3) the cardinal value model, and (4) the response surface models, or polynomial models.

The *Arrhenius models* are based on the empirical expression that relates the growth rate to the environmental temperature. The basis for this expression was taken from the work of van't Hoff and Arrhenius from the end of the nineteenth century examining the thermodynamics of chemical reactions.

$$k = Ae^{-E_a/RT}, \tag{3.11}$$

where k is the growth rate (h^{-1}), A is the "collision factor" (h^{-1}), E_a is "activation energy" of the reaction system ($kJ\ mol^{-1}$), R is the universal gas constant ($8.31\ kJ\ mol^{-1}\ K^{-1}$), and T is the absolute temperature (K)

The Arrhenius equation was extended by Davey (1989). Davey modeled the growth rate, here represented by k, as a function of temperature and water activity. The factors C_0–C_4 are the parameters to be modeled. When water activity is not a limiting value, the last terms can be removed:

$$\ln k = C_0 + \frac{C_1}{T} + \frac{C_2}{T^2} + C_3 a_w + C_4 a_w^2. \tag{3.12}$$

A second group of secondary models consists of the family of *Belehradek models*, or *square-root models*. The first attempt was made by Ratkowsky et al. (1982), who suggested a simple relationship between the growth rate k and temperature. The equation was only to be applied to the low-temperature region:

$$\sqrt{k} = b(T - T_{min}). \tag{3.13}$$

T_{min} is an estimated extrapolation of the regression line derived from a plot of $\sqrt{k}$ as a function of temperature. By definition, the growth rate at this point is zero. It should be stressed that T_{min} is usually 2–3°C lower than the temperature at which growth is usually observed, so T_{min} has to be interpreted as a conceptual temperature (McMeekin et al., 1993).

This basic square-root model has been adjusted several times by several authors. Some of the adaptations will be highlighted here, without the aim of being complete. A first adaptation was made by Ratkowsky et al. (1983), enlarging the scope to the whole temperature range:

$$\sqrt{k} = b(T - T_{min})\left\{1 - \exp\left[c(T - T_{max})\right]\right\}. \tag{3.14}$$

Other adaptations aimed to include growth factors other than only temperature. McMeekin et al. (1987) added an a_w parameter to the basic model, while Adams et al. (1991) included a pH term. McMeekin et al. (1992) suggested that the two additional factors could be included at the same time. Later on, Devlieghere et al. (1999) replaced the pH terms with terms of dissolved CO_2 to describe the behavior of spoilage flora in modified-atmosphere packaged meat products. The main form as described by McMeekin et al. (1992) is:

$$\sqrt{k} = b\sqrt{(a_w - a_{w,min})}\sqrt{(pH - pH_{min})}(T - T_{min}).$$ (3.15)

A third group of secondary models is formed by the *cardinal models*. The first model was developed by Rosso et al. (1995). It describes the effect of temperature and pH on the growth rate of a microorganism, based on the cardinal values: the optimal, the minimal, and the maximal temperature and pH at which growth is possible:

$$\mu_{max}(T, pH) = \mu_{opt} \tau(T) \rho(pH),$$ (3.16)

where

$$T < T_{min}, 0,$$
$$\tau(T) = T_{min} < T < T_{max},$$
$$\frac{(T - T_{max})(T - T_{min})^2}{(T_{opt} - T_{min})\left[(T_{opt} - T_{min})(T - T_{opt}) - (T_{opt} - T_{max})(T_{opt} + T_{min} - 2T)\right]},$$
$$T > T_{max}, 0,$$
$$pH < pH, 0,$$

$$\rho(pH) = pH_{min} < pH < pH_{max}, \frac{(pH - pH_{min})(pH - pH_{max})}{(pH - pH_{min})(pH - pH_{max}) - (pH - pH_{opt})^2},$$

and

$$pH > pH_{max}, 0$$

All parameters used within this model have a biological interpretation, although the model structure itself is a pure empirical relationship.

A fourth group of secondary models consists of the *response surface models*, or *polynomial models*. Those are pure empirical black box models, where fitting capacities are the most beneficial properties, and the model structure is easy and easy to fit. Classically, a second-order polynomial function is used, so the model will consist of three groups of terms: the first-order terms, the second-order terms, and the interaction terms. Here an example is given, a second-order model describing the growth rate as a function of temperature (T), concentration of dissolved CO_2 (C), a_w, and sodium lactate concentration (L) (Devlieghere et al., 2000):

$$\mu = I_\mu + m_1 T + m_2 a_w + m_3 L + m_4 C_2 + m_5 T^2 + m_6 a_w^2 +$$
$$m_7 L^2 + m_8 C^2 + m_9 T a_w + m_{10} TL + m_{11} TC +$$
$$m_{12} a_w L + m_{13} a_w C + m_{14} CL.$$ (3.17)

When working with polyomial models, one has to estimate a lot of parameters from the data, and the number increases exponentially when more environmental factors are added to the model. Not all of these parameters are statistically relevant, so often a backward regression is executed to eliminate the nonrelevant parameters.

Sometimes, these models are sneering called "*bulldozer models*" because of the high computational capacities that are needed to estimate the high number of model parameters and the total absence of a mechanistic approach. Although less elegant, the polynomial models can result in very good fitting capacities and are neverthe-less very often used in practice.

3.2.3.3   Tertiary Models

Tertiary models are software packages in which the previously described models are integrated into a ready-to-use application tool. As such, the discipline of predic-tive modeling can find its way to the food industry.

The most well-known tertiary model is the Pathogen Modeling Program (PMP) (Fig. 3.3), which was developed by the USDA in the 1990s. It describes the behavior of several pathogenic bacteria as a function of the environmental conditions. The model contains growth curves, inactivation curves, and survival, cooling and irradiance mod-els. It also makes predictions for the time to turbidity or the time to toxin production of

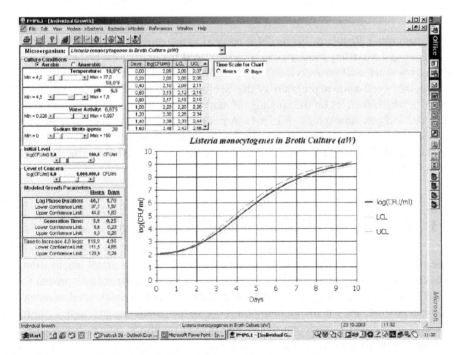

Fig. 3.3   Screen print from the Pathogen Modeling Program model

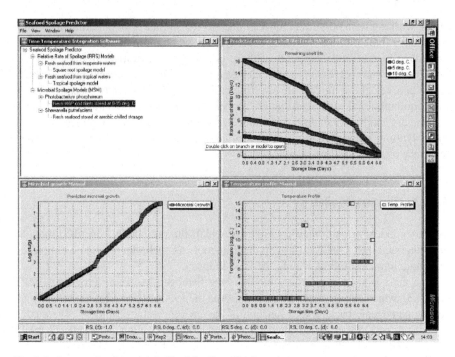

Fig. 3.4 Screen print from the Seafood Spoilage Predictor

some pathogens. Regularly, the model is updated to the current modeling knowledge. The Pathogen Modeling Program can be downloaded freely from the USDA Web site (http://www.ars.usad.gov/services/software/download.htm?softwareid=90).

Another well-known program is the Seafood Spoilage and Safety Predictor (SSSP), which focuses on the spoilage of marine, fresh fish, as a function of temperature and gas atmosphere (Fig. 3.4). A realistic time–temperature combination can be defined and the subsequent microbial growth is predicted, together with the remaining shelf life. The program can be freely downloaded from the Internet (http://www.dfu.min.dk/micro/ssp).

3.3 Modeling the Inactivation of Microorganisms

The modeling of inactivation, more specifically modeling of thermal inactivation, was one of the first applications of the discipline that was later called "predictive microbiology." Apart from the thermal inactivation, also the nonthermal inactivation can be modeled. Chemical inhibition is one of the recent research topics within the area of predictive modeling. Other nonthermal inhibition factors can be high pressure or irradiation.

3.3.1 *Modeling Thermal Inactivation*

Classically, the *D* and *Z* values are used to model the thermal resistance of a micro-organism. When the cell count is pictured as a semilogarithmic graph with the logarithm of the cell count on the *Y*-axis and time on the *X*-axis (Fig. 3.5), the thermal reduction at a certain temperature has a linear relationship. *D* is then defined as the time needed to obtain one decimal reduction of the cell count.

When the *D* is determined at several temperatures, *Z* can be calculated. When *D* is depicted as a function of temperature on a semilogarithmic scale, *Z* can be defined as the temperature rise that is needed to decrease *D* by a factor 10 (Fig. 3.6).

The linear relationship between the cell count and the heating time is an ideal situation. In practice, there can be some shoulder or tailing effects, especially when dealing with low heating temperatures. The shoulder effects can be explained by a

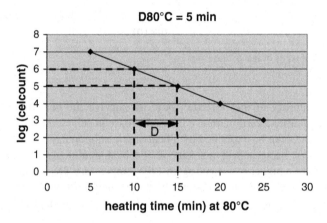

Fig. 3.5 *D* value

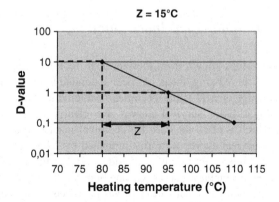

Fig. 3.6 *Z* value

Inactivation curves

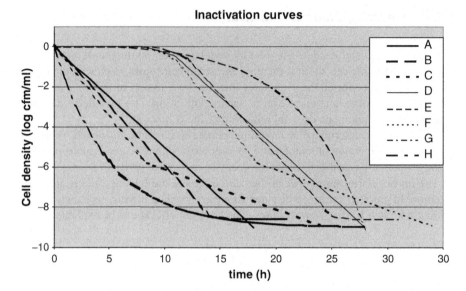

Fig. 3.7 Different types of inactivation curves

limited heat stability of the bacteria: they can survive the heat treatment for a short period, but after a certain time, a linear decrease in cell count can be observed. The tailing effects can be explained by a heat-stable subpopulation within the total population. Most bacteria will not survive the heat treatment, although a small group might have a higher resistance against the temperature stress, causing a tailing effect in the curve. More information about the modeling of these inactivation curves can be found in Sect. 3.3.2, where the different curve-shapes are discussed (Fig. 3.7).

3.3.2   Modeling Nonthermal Inactivation

The concepts of D values known from thermal heat treatments, assuming a semi-logarithmic survival curve being linear, has been extended to chemical inactivation. Obviously, the concept of a D value becomes problematic when experimentally determined semilogarithmic survival curves are clearly nonlinear (Peleg and Penchina, 2000) which is often the case for chemical inactivation. Six commonly observed types of survival curves were distinguished by Xiong et al. (1999): log-linear curves (A), log-linear curves with a shoulder (D), log-linear curves with a tail (B), sigmoidal curves (E), biphasic curves (C), and biphasic curves with a shoulder (F). Two other shapes of chemical inactivation curves have been reported by Peleg and Penchina (2000), being concave downward (G) and concave upward (H) curves. The eight different inactivation curves are shown in Fig. 3.7.

Most of the attention of model development in the field of chemical inactivation has focused on development of primary models, and to a much less extent on secondary model development. In most cases the influence of environmental factors of the microbial cell on the primary model parameters are described by black box polynomial equations. In many cases the time for a 4D reduction (t_{4D}) is described by a polynomial equation as a function of factors, including temperature, pH, and concentration of the chemical(s) applied. This was, for example, the case for an expanded model for nonthermal inactivation of *Listeria monocytogenes* (Buchanan et al., 1997b). The effect of temperature, salt concentration, nitrite concentration, lactic acid concentration, and pH on the inactivation of *Listeria monocytogenes* was examined:

$$\ln(t_{4D}) = 0.0371T + 0.0575S + 3.902P - 1.749L - 0.0547N - 0.0012TS$$
$$- 0.00812TP - 0.0131TL + 0.000323TN + 0.0103SP + 0.00895SL$$
$$+ 0.0001582SN + 0.1895PL + 0.00356PN + 0.00209LN$$
$$- 0.00168T^2 - 0.00749S^2 - 0.3007P^2 + 0.1705L^2 + 0.0000871N^2$$
$$- 3.666, \tag{3.18}$$

where T is temperature (4–42°C), S is sodium chloride concentration (0.5–19.0% aqueous phase), P is pH (3.2–7.3), L is lactic acid concentration (0–2% w/w), and N is sodium nitrite concentration (0–200 μg/ml).

3.3.3   Modeling the Growth/No Growth Interface

At the interface between the growth models and the inactivation models there is a growth/no growth twilight zone. Many large food producers choose nongrowth supporting conditions to guarantee the microbial safety of their food product, i.e., they do not allow any growth of any food pathogen in the product. For them it is essential to know the interface between conditions supporting growth and conditions where growth is not possible. This interface can be defined by one single factor such as temperature (cardinal value) when other factors remain constant but more often a combination of factors such as temperature, pH, a_w, and concentrations of a chemical compound is considered. Several models have been developed to describe these multiple-factor interfaces.

When the growth/no growth interface is modeled, a lot of data near the growth boundaries have to be collected, as the abruptness of the transition between growth and no growth is often quite steep. Examples can be found in Tienungoon et al. (2000), where a pH decrease of 0.1–0.2 units was reported to cause cessation of growth for *Listeria monocytogenes*, while Salter et al. (2000) showed that an a_w decrease of 0.001–0.004 forms the growth/no growth boundary for *Escherichia coli*.

A first attempt to model the growth/no growth interface was made by Ratkowsky and Ross (1995). They transformed the square-root model (*a kinetic model*) of McMeekin et al. (1992), adjusted with a nitrite factor *(N)* (Eq. 3.19), to a logit-based model (*a probability model*) (Eq. 3.20):

$$\sqrt{\mu_{max}} = a_1(T - T_{min})\sqrt{(pH - pH_{min})(a_w - a_{w,min})(N_{max} - N)}. \quad (3.19)$$

From both sides of the equation, the natural logarithm was taken, and the left side was replaced by $logit(p)=\ln[p/(1-p)]$, where p is the probability that growth occurs. The dataset is considered to be binary: $p = 1$ if growth can be observed for that defined combination of parameters, while $p = 0$ if no growth is observed. The new equation is

$$\begin{aligned} logit(p) = b_0 + b_1 \ln(T - T_{min}) + b_2 \ln(pH - pH_{min}) \\ + b_3 \ln(a_w - a_{w,min}) + b_4 \ln(N_{max} - N) \end{aligned} \quad (3.20)$$

The coefficients b_0, b_1, b_2, b_3, and b_4 have to be estimated by fitting the model to experimental data. The other parameters (T_{min}, pH_{min}, $a_{w,min}$, and NO_{2max}) should be estimated independently or fixed to constant values, so the coefficients can be estimated using linear logistic regression. Once the b values have been predicted, the position of the interface can be estimated by choosing a fixed p. Often $p = 0.5$ is preferred, corresponding to a 50:50 chance that the organism will grow under those fixed environmental conditions. Then $logit(0.5) = \ln(0.5/0.5) = \ln(1) = 0$.

A second type of model was developed by Masana and Baranyi (2000). They investigated the growth/no growth interface of *Brochotrix thermosphacta* as a function of pH and water activity. The a_w values were recalculated as $b_w = \sqrt{(1 - a_w)}$. The model could be divided into two parts: a parabolic part 1 and a linear part at a constant NaCl level (Eq. 3.21):

$$pH_{boundary} = a_0 + a_1 b_w + a_2 b_w^2. \quad (3.21)$$

A third type of modeling was proposed by Le Marc et al. (2002). The model, describing the *Listeria monocytogenes* growth/no growth interface, is based on a four-factor kinetic model estimating the value of μ_{max}. The functions are based on the cardinal models of Rosso et al. (1995):

$$\mu_{max} = \mu_{opt}\rho(T)\gamma(pH)\tau([RCOOH])\xi(T, pH, [RCOOH]). \quad (3.22)$$

The growth/no growth interface was obtained by reducing μ_{max} to zero. This can be done by making the interaction term $\xi(T,pH,[RCOOH])$ equal to zero.

3.4   Model Development Procedures

The development of a predictive model in the field of food microbiology is basically performed in four steps: (1) planning, (2) data collection and analysis, (3) function fitting, and (4) validation and maintenance (McMeekin et al., 1993). Each step is essential in the model development (Fig. 3.8).

In the planning step, it is essential to know the requirements of the final user of the model. The planning will differ when a model is developed to estimate the direct effect and interactions of independent variables on the growth of a microorganism compared with that for a model that is developed to predict quantitatively the growth of the microorganism in a specific food product. The decisions to be made in the planning step are summarized in Table 3.1.

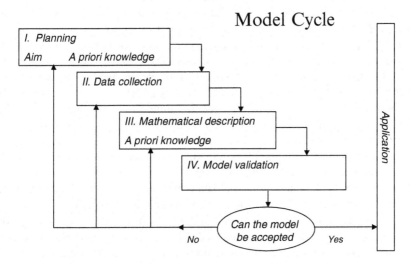

Fig. 3.8   The model cycle

Table 3.1   Important questions for the planning step in model development

What is the objective?

What are the most important independent variables for this objective?

What is the (practical) range of these independent variables?

What type of inoculum is going to be applied (physiological status, inoculum size, mixture or single strain, way of inoculation)?

Is a background flora important for the defined objective?

What dependent variables are going to be determined?

What substrate is going to be used?

How much and which combinations of the independent variables will be incorporated in the experimental design?

How will the data be collected?

Preliminary experiments are often of major importance for achieving the postulated objective with a minimum of time and resources. Reduction of the number of independent variables incorporated in the developed model, for example, can lead to very significant reductions of labor and material costs. Adequate predictive models are indeed difficult to develop when more than four independent variables have to be included. It is therefore necessary to determine the relevant intrinsic and extrinsic parameters for a specific food product and to develop specific food-related predictive models based on this knowledge (Devlieghere et al., 2000). The determination of the relevant independent variables can be based on present knowledge combined with preliminary experiments.

The development of a model requires the collection of data from experimentation. Growth of microorganisms can be followed by different enumeration methods. *Viable count methods* are generally applied. The advantages of plate counting methods are that they can be made very sensitive, they measure only viable cells, and they can be used to determine microbial loads in both real foods and laboratory liquid media (McMeekin et al., 1993). Their chief disadvantage, however, is the amount of labor and materials required to obtain a single determination. A number of workers have therefore chosen to construct their initial models using data obtained from *optical density determinations* in laboratory media. The advantages of this method are speed, simplicity, and noninvasiveness, but many disadvantages have been reported: limitation to clear media, limited range of linearity between concentration and absorbance/turbidity, low detection limit (10^6 cells/ml), low upper limit ($10^{7.5}$ cells/ml) of the linear part, and lag determinations can only be made on dense populations (McMeekin et al., 1993), and there is a higher risk for misinterpretation (McDonald and Sun, 1999).

Optimization of the quantity of the data collected and the costs involved with data collection is necessary. To fit primary models and subsequent secondary models, data must be collected over the entire growth period and often over 100 primary growth curves have to be produced (McDonald and Sun, 1999). Bratchell et al. (1989) suggested that 15–20 observations per growth curve are sufficient, but using only ten can lead to large differences between surfaces fitted with data obtained with high and low numbers of observations per growth curve. Gibson et al. (1987) stated that up to ten to 15 data points are required to obtain growth curves, which can be modeled reliably by sigmoidal growth models. Poschet and Van Impe (1999) concluded from their study, in which the influence of data points on the uncertainty of a model was analyzed by means of Monte Carlo analysis, that the uncertainty on the model parameters decreases with increasing number of data points. However, from a certain number of data points on, the uncertainty did not decrease anymore with increasing number of data points.

The data obtained are fitted to one of the existing primary models expressing the evolution of the cell density as a function of time (i.e., growth curves, see Sect. 3.2.3.1) to obtain growth parameters (such as maximum specific growth rate, lag phase) for each combination of the environmental conditions. In a second step, the combinations of environmental factors obtained with their respective growth parameters are fitted to a secondary model function as described in Sect. 3.2.3.2.

Function fitting is performed by regression analysis that can be linear or nonlinear, depending on the character of the model applied. The actual process of fitting a function to a dataset, i.e., to determine the parameter values that best fit the model to those data, is based on the principle of least squares. This criterion aims to derive parameter values that minimize the sum of the squares of the differences between observed values and those predicted by the fitted model, i.e., the residuals (McMeekin et al., 1993).

Predictive microbiology aims at the quantitative estimation of microbial growth in foods using mathematical modeling. To determine whether predictions provide good description of growth in foods, models should be validated to evaluate their predictive ability (te Giffel and Zwietering, 1999). Validation should be performed in media, but especially in real food products.

The accuracy of models can be assessed graphically by plotting the observed values against corresponding predictions of a model. Furthermore, mean square error and R^2 values can also be used as an indication of the reliability of models when applied in foods. Ross (1996) proposed indices for the performance of mathematical models used in predictive microbiology. The objective of those performance indices was to enable the assessment of the reliability of such models when compared with observations *not* used to generate the model, particularly in foods, and hence to evaluate their utility to assist in food safety and quality decisions (Baranyi et al., 1999). A further objective was to provide a simple and quantitative measure of model reliability.

Those indices were termed bias factor (B_f) and accuracy factor (A_f) and are calculated as follows:

$$B_f = 10^{\left(\sum \log(GT_{predicted}/GT_{observed})/n\right)},$$
(3.23)

$$A_f = 10^{\left(\sum |\log(GT_{predicted}/GT_{observed})|/n\right)}.$$
(3.24)

B_f and A_f defined by Ross (1996) were refined by basing the calculation of those measures on the mean square differences between predictions and observations (Baranyi et al., 1999). Here the most general form is given by comparing the accuracy factor and the bias factor for models f and g. When comparing a model with the observed values, the prediction values of model g are replaced by the observed values:

$$B_{f,g} = \exp\left(\frac{\int_R [\ln f(x_1,...,x_n) - \ln g(x_1,...,x_n)] dx_1...dx_n}{V(R)}\right),$$
(3.25)

$$A_{f,g} = \exp\left(\sqrt{\frac{\int_R [\ln f(x_1,...,x_n) - \ln g(x_1,...,x_n)]^2 dx_1...dx_n}{V(R)}}\right),$$
(3.26)

where $f(x_1,...,x_n)$ is the predicted value of the growth rate μ as a function f determined by n environmental parameters x, $g(x_1,...,x_n)$ is the predicted value of the growth rate μ as a function g determined by n environmental parameters x, and $V(R)$ is the volume of the region $\int_R 1\,dx_1,...,dx_n$, e.g., if R is a temperature interval (T_1-T_2), then $V(R) = T_2-T_1$.

3.5 Current Deficiencies in Predictive Microbiology

3.5.1 Interactions Between Microorganisms

Competition between microorganisms in foods is not considered in most predictive models and very few have been published regarding modeling of microbial interactions in food products. Recently, more attention has been given to incorporating interactions in predictive models, especially describing the effect of lactic acid bacteria on the behavior of food-borne pathogens.

Vereecken and Van Impe (2002) developed a model for the combined growth and metabolite production of lactic acid bacteria, a microbial group that is often used as a starter culture for fermented foods, such as dairy products or fermented meat products. Within the food product, the lactic acid bacteria are used as a protective culture – or biopreservative – against pathogens and spoilage organisms. The model of Vereecken and Van Impe is based on the production of lactic acid by the lactic acid bacteria. The transition from the exponential phase to the stationary phase, which is the growth inhibition, could be modeled by the concentration of undissociated lactic acid and the acidification of the environment. Later, a co-culture model was proposed by Vereecken et al. (2003), based on the same theory, describing the mono- and co-culture growth of *Yersinia enterocolitica*, a pathogen, together with *Lactobacillus sakei*, a lactic acid bacterium. The inhibition of the pathogen could be modeled by the lactic acid production of the *Lactobacillus sakei*.

3.5.2 Structure of Foods

During the last decade, a large variety of predictive models for microbial growth have been developed. Most of these models do not take into account the variability of microbial growth with respect to space. In homogeneous environments, such as broths and fluid foods, this variability does not exist or may be neglected (Vereecken et al., 1998). However, in structured (heterogeneous) foods, growth determinative factors such as temperature and pH and thus microbial growth itself may be highly related to the position in the food product. Moreover, owing to the solid structure of the food, microorganisms can be forced to grow in colonies, in which competition and interaction effects play a more important role in comparison with fluid products.

Dens and Van Impe (2000) applied a coupled map lattice approach to simulate the spatiotemporal behavior of a mixed contamination of *Lactobacillus plantarum* and *Escherichia coli* on a food with a very low diffusion coefficient, which is the case in solid foods. They demonstrated pattern formation, making global coexistence of both species possible, which was not the case in a homogeneous liquid medium. Recently the model was extended (Dens and Van Impe, 2001) to describe two phenomena: (1) the local evolution of the biomass and (2) the transfer of biomass trough the medium. In this situation, predictive models based on experiments in broth overestimate the safety of a solid food product.

3.5.3   Diversity of Natural Strains

Naturally occurring strains of food pathogens as well as spoilage microorganism differ in their response to stress conditions. Begot et al. (1997) compared the growth parameters of 58 *Listeria monocytogenes* and eight *Listeria innocua* strains at different combinations of pH (5.6 and 7.0), temperature (10 and 37°C), and water activity (0.96 and 1.00). Wide variations between the different strains were observed at the same conditions, especially for the lag phase. At one combination ($a_w = 0.96$; $T = 10°C$; pH 7.0) the minimum lag phase was 3.9 h, while the maximum lag phase amounted to 97.9 h. In the same conditions, the maximum generation time was 2.3 times the minimum generation time. Moreover, strains exhibiting the longest generation time of the entire population did not show the maximal lag time. By cluster analysis, it was demonstrated that the group of the fastest-growing strains at all conditions was mainly composed of strains isolated from industrial sites. Such results point out the difficulties in choosing a strain to build a predictive model. If the fastest strain is considered, the model will always predict a shorter lag time and a lower growth rate than those found for the majority of strains, and would thus always provide safe growth estimates. On the other hand, questions could arise about the reality of the predictions made by models developed with the fastest-growing strain. The ideal model should also quantify the uncertainty of the predictions, caused by the diversity of natural strains.

3.5.4   Modeling of the Lag Phase

The variability of a population lag phase, i.e., the time needed for a bacterium to adapt to a new environment before it starts multiplying, is an important source of variability in the exposure assessment step when performing a risk assessment concerning *Listeria monocytogenes*. Research on modeling the behavior of bacteria in foods has repeatedly shown that the lag phase is more difficult to predict than the specific growth rate.

Preadaptation to inimical growth conditions can shorten lag times dramatically and the magnitude of this effect is difficult to predict. Also the inoculum level has an important impact on the lag phase. It has been observed that the variability of detection times of *Listeria monocytogenes* increases when a lower inoculum level is applied and, when one is dealing with constant inoculum levels, an increase in variation can be observed when more salt stress is applied (Robinson et al., 2001). Therefore, when dealing with realistic low inoculum levels of pathogens such as, e.g., *Listeria monocytogenes*, a high variability can be expected in the population lag phase, especially when severe stress conditions are applied. Higher initial densities of bacteria are theoretically associated with a higher likelihood of including at least one cell in the proper physiological state for immediate growth, with only a limited lag time for adjustment (Baranyi, 1998). An individual cell approach is therefore needed to adequately quantify these distributions.

Even when inoculum effects have been minimized, it has still proved difficult to obtain a clear picture of the way lag time varies as a function of the external environment. Several studies have demonstrated a relationship between lag time and growth rate, but the general validity of this relationship has not been fully explored. A clear understanding about the mechanisms of action of stress conditions on the microbial cell as well as the cellular adaptation systems will probably lead to better predictions of the lag phase. Until then, however, the lag phase should not be excluded for prediction of spoilage or microbial safety of food products, but careful interpretation of the model predictions is necessary.

3.5.5 Modeling Behavior of Spoilage Microorganisms

As the criteria for the shelf life of most food products are set by a combination of criteria for food pathogens and for spoilage microorganisms, mathematical models to predict behavior of the major groups of spoilage microorganisms would be useful. Development of comprehensive models for spoilage organisms has not received much attention. The development of mathematical models to predict microbial spoilage of foods is normally very product- and/or industry-specific, which limits the development of such predictive models. However, in recent years research into predicting spoilage has gained more interest.

3.5.6 Dealing with Unrealistic Modeling Predictions

When working with predictions, one sometimes makes unrealistic predictions. This is mostly the case for black box models, containing no mechanistic elements and having unrealistic high R^2 values but impossible results as unrealistic curvatures or

Table 3.2 Possible applications of predictive food microbiology

Hazard analysis critical control point
  Preliminary hazard analysis
  Identification and establishment of critical control point
  Corrective actions
  Assessment of importance of interaction between variables
Risk assessment
  Estimation of changes in microbial numbers in a production chain
  Assessment of exposure to a particular pathogen
Microbial shelf life studies
  Prediction of the growth of specific food spoilers
  Prediction of growth of specific food pathogens
Product research and development
  Effect of altering product composition on food safety and spoilage
  Effect of processing on food safety and spoilage
  Evaluation of effect of out-of specification circumstances
Temperature function integration and hygiene regulatory activity
  Consequences of temperature in the cold chain for safety and spoilage
Education
  Education of technical and especially nontechnical people
Design of experiments
  Number of samples to be prepared
  Defining the interval between sampling

even negative lag phases or negative growth rates can occur. Geeraerd et al. (2004) developed a novel approach for secondary models incorporating microbial knowledge in black box polynomial models. By inserting some a priori microbial knowledge into the black box models, the goodness of fit might be a little lower, but the model will be more rigid and realistic model outputs can be expected.

3.6 Applications of Predictive Microbiology

Use of predictive models can quickly provide information and, therefore, it is important to appreciate the real value and usefulness of predictive models. It is, however, important to point out that their applications cannot completely replace microbial analysis of samples or the sound technical experience and judgment of a trained microbiologist (McDonald and Sun, 1999). It can be described by a quote from te Giffel and Zwietering (1999): "A model is a useful 'discussion partner' giving you good ideas, pointing you in the right direction, but like other discussion partners is not always right." It has to be considered as one of the tools decision makers, active in the field of food, have at their disposal to consolidate their decisions. Many applications, which are summarized in Table 3.2, have been proposed for predictive food microbiology.

References

Adams, M.R., Little, C.L. and Easter, M.C. Modeling the effect of pH, acidulant and temperature on the growth rate of *Yersinia enterocolitica*, *J. Appl. Bacteriol.*, 71, 65–71, 1991.

Baranyi, J., Comparison of statistic and deterministic concepts of bacterial lag, *J. Theor. Biol.*, 192, 403–408, 1998.

Baranyi, J. and Roberts, T.A., A dynamic approach to predicting bacterial growth in food, *Int. J. Food Microbiol.*, 23, 277–294, 1994.

Baranyi, J., Pin, C. and Ross, T., Validating and comparing predictive models, *Int. J. Food Microbiol.*, 48, 159–166, 1999.

Begot, C., Lebert, I. and Lebert, A., Variability of the response of 66 *Listeria monocytogenes* and *Listeria innocua* strains to different growth conditions, *Food Microbiol.*, 14, 403–412, 1997.

Bratchell, N., Gibson, A.M., Truman, M., Kelly, T.M. and Roberts, T.A., Predicting microbial growth: the consequences of quantity of data, *Int. J. Food Microbiol.*, 8, 47–58, 1989.

Buchanan, R.L., Whiting, R.C. and Damert, W.C., When simple is good enough: a comparison of the Gompertz, Baranyi, and three-phase linear models for fitting bacterial growth curves, *Food Microbiol.*, 14, 313–326, 1997a.

Buchanan, R.L., Golden, M.H. and Phillips, J.G., Expanded models for the non-thermal inactivation of Listeria monocytogenes, *J. Appl. Microbiol.*, 82, 567–577, 1997b.

Davey, K.R., A predictive model for combined temperature and water activity on the microbial growth during the growth phase, *J. Appl. Bacteriol.*, 67, 483–488, 1989.

Dens, E.J. and Van Impe, J.F., On the importance of taking space into account when modeling microbial competition in structured foods. *Math. Comput. Simulat.*, 53 (4–6), 443–448, 2000.

Dens, E.J. and Van Impe, J.F., On the need for another type of predictive model in structured foods, *Int. J. Food Microbiol.*, 64, 247–260, 2001.

Devlieghere, F., Van Belle, B. and Debevere, J., Shelf life of modified atmosphere packed cooked meat products: a predictive model, *Int. J. Food Microbiol.*, 46, 57–70, 1999.

Devlieghere, F., Geeraerd, A.H., Versyck, K.J., Bernaert, H., Van Impe, J. and Debevere, J., Shelf life of modified atmosphere packaged cooked meat products: addition of Na-lactate as a fourth shelf life determinative factor in a model and product validation. *Int. J. Food Microbiol.*, 58, 93–106, 2000.

Geeraerd, A.H., Valdramidis, V.P., Devlieghere, F., Bernaert, H., Debevere, J. and Van Impe, J.F., Development of a novel approach for secondary modeling in predictive microbiology: incorporation of microbiological knowledge in black box polynomial modeling, *Int. J. Food Microbiol.*, 91 (3), 229–244, 2004.

Gibson, A.M. and Hocking, A.D., Advances in the predictive modeling of fungal growth, *Trends Food Sci. Technol.*, 8 (11), 353–358, 1997.

Gibson, A.M., Bratchell, N. and Roberts, T.A., The effect of sodium chloride and temperature on the rate and extent of growth of *Clostridium botulinum* type A in pasteurized pork slurry, *J. Appl. Bacteriol.*, 62, 479–490, 1987.

Le Marc, Y., Huchet, V., Bourgeois, C.M., Guyonnet, J.P., Mafart, P. and Thuault, D., Modeling the growth kinetics of *Listeria monocytogenes* as a function of temperature, pH and organic acid concentration, *Int. J. Food Microbiol.*, 73, 219–237, 2002.

Masana, M.O. and Baranyi, J., Growth/no growth interface of *Brochotrix thermosphacta* as a function of pH and water activity, *Food Microbiol.*, 17, 485–493, 2000.

McDonald, K. and Sun, D.-W., Predictive food microbiology for the meat industry: a review, *Int. J. Food Microbiol.*, 53, 1–27, 1999.

McMeekin, T.A., Chandler, R.E., Doe, P.E., Garland, C.D., Olley, J., Putro, S. and Ratkowsky, D.A., Model for the combined effect of temperature and water activity on the growth rate of *Staphylococcus xylosus*, *J. Appl. Bacteriol.*, 62, 543–550, 1987.

McMeekin, T.A., Ross, T. and Olley, J., Application of predictive microbiology to assure the quality and safety of fish and fish products, *Int. J. Food Microbiol.*, 15, 13–32, 1992.

McMeekin, T.A., Olley, J.N., Ross, T. and Ratkowsky, D.A., Predictive microbiology – theory and application, Wiley, New York, 340 p, 1993.

Peleg, M. and Penchina, C.M., Modeling microbial survival during exposure to a lethal agent with varying intensity, *Crit. Rev. Food Sci. Nutr.*, 40 (2), 159–172, 2000.

Poschet, F. and Van Impe, J.F., Quantifying the uncertainty of model outputs in predictive microbiology: a Monte Carlo analysis, *Med. Fac. Landbouwwet. Universiteit Gent.*, 64 (5), 499–506, 1999.

Ratkowsky, D.A. and Ross, T., Modeling the bacterial growth/no growth interface, *Lett. Appl. Microbiol.*, 20, 29–33, 1995.

Ratkowsky, D.A., Olley, J., McMeekin, T.A. and Ball, A., Relationship between temperature and growth rate of bacterial cultures, *J. Bacteriol.*, 149 (1), 1–5, 1982.

Ratkowsky, D.A., Lowry, R.K., McMeekin, T.A., Stokes, A.N. and Chandler, R.E., Model for bacterial culture growth rate throughout the entire biokinetic temperature range, *J. Bacteriol.*, 154 (3), 1222–1226, 1983.

Robinson, T.P., Aboaba, O.O., Ocio, M.J., Baranyi, J. and Mackey, B.M., The effect of inoculum size on the lag phase of *Listeria monocytogenes*, *Int. J. Food Microbiol.*, 70, 163–173, 2001.

Ross, T., Indices for performance evaluation of predictive models in food microbiology, *J. Appl. Bacteriol.*, 81, 501–508, 1996.

Ross, T. and McMeekin, T.A., Predictive microbiology, *Int. J. Food Microbiol.*, 23, 241–264, 1994.

Ross, T., Dalgaard, P. and Tienungoon, S., Predictive modeling of the growth and survival of *Listeria* in fishery products, *Int. J. Food Microbiol.*, 62, 231–245, 2000.

Rosso, L., Lobry, J.R., Bajard, S. and Flandrois, J.P., Convenient model to describe the combined effects of temperature and pH on microbial growth, *Appl. Environ. Microbiol.*, 61 (2), 610–616, 1995.

Salter, M.A., Ratkowsky, D.A., Ross, T. and McMeekin, T.A., Modeling the combined temperature and salt (NaCl) limits for growth of a pathogenic *Escherichia coli* strain using nonlinear logistic regression, *Int. J. Food Microbiol.*, 61, 159–167, 2000.

te Giffel, M.C. and Zwietering, M.H., Validation of predictive models describing the growth of *Listeria monocytogenes*, *Int. J. Food Microbiol.*, 46, 135–149, 1999.

Tienungoon, S., Ratkowsky, D.A., McMeekin, T.A. and Ross, T. Growth limits of *Listeria monocytogenes* as a function of temperature, pH, NaCl and Lactic acid, *Appl. Environ. Microbiol.*, 66 (11), 4979–4987, 2000.

Van Impe, J.F., Nicolaï, B.M., Martens, T., De Baerdemaeker, J. and Vandewalle, J., Dynamic mathematical model to predict microbial growth and inactivation during food processing, *Appl. Environ. Microbiol.*, 58 (9), 2901–2909, 1992.

Vereecken, K.M. and Van Impe, J.F., Analysis and practical implementation of a model for combined growth and metabolite production of lactic acid bacteria, *Int. J. Food Microbiol.*, 73, 239–250, 2002.

Vereecken, K., Bernaerts, K., Boelen, T., Dens, E., Geeraerd, A., Versyck, K. and Van Impe, J., State of the art in predictive food microbiology, *Med. Fac. Landbouwwet. Universiteit Gent.*, 63 (4), 1429–1437, 1998.

Vereecken, K.M., Devlieghere, F., Bockstaele, A., Debevere, J. and Van Impe, J.F., A model for lactic acid-induced inhibition of *Yersinia enterocolitica* in mono- and coculture with *Lactobacillus sakei*, *Food Microbiol.*, 20, 701–713, 2003.

Whiting, R.C. and Buchanan, R.L., A classification of models in predictive microbiology – reply, *Food Microbiol.*, 10 (2), 175–177, 1993.

Whiting, R.C. and Buchanan, R.L., Microbial modeling, *Food Technol.*, 48 (6), 113–120, 1994.

Xiong, R., Xie, G., Edmondson, A.E. and Sheard, M.A., A mathematical model for bacterial inactivation, *Int. J. Food Microbiol.*, 46, 45–55, 1999.

Chapter 4
Prediction of Heat Transfer During Food Chilling, Freezing, Thawing, and Distribution

Christian James, Laurence Ketteringham, Silvia Palpacelli, and Stephen James

4.1 Introduction

Food engineers need to know the amount of time it takes to freeze, chill, or thaw their products and how much energy is required for that process. They also need to know the effect of storage, transport, or display conditions on the product temperatures. Currently experience plays a large role in identifying the time and energy needed, especially in situations where the food products can change from one day to the next. A reliable mathematical model can be of considerable help in optimizing a process and investigating the consequences of design changes.

4.2 Prediction Methods

A number of predictive modeling techniques, usually in combination with limited experimentation, have been used over the years to generate data and optimize refrigeration processes.

4.2.1 Empirical Methods

Empirical models can be generated by carrying out a range of experiments in pilot plants where processing conditions, temperature, velocity, humidity, etc. can be accurately controlled. The cooling characteristics of different combinations of extrinsic and intrinsic conditions can then be accessed and used to predict what effect other conditions may have on cooling characteristics. Generally, one experiment is carried out under constant conditions, but the refrigerating media properties are varied from one trial to another. The weight, grade, etc. of the product are also varied to cover the usual range of practical cooling conditions. For example, the

C. James (✉)
Food Refrigeration and Process Engineering Research Centre, University of Bristol,
Churchill Building, Langford, Bristol BS40 5DU, UK

R. Costa, K. Kristbergsson (eds.), *Predictive Modeling and Risk Assessment*,   55
DOI: 10.1007/978-1-387-68776-6, © Springer Science + Business Media, LLC 2009

data may show, for pork hindquarters, that chilling time increases by 35% when the carcass weight varies from 50 to 100 kg and by 25% when the air velocity increases from 0.5 to 2 m s^{-1} (Daudin and van Gerwen, 1996).

Chilling rate curves can be simply turned into models using Newton's law of cooling. The chilling rate is the slope of the linear part of the logarithmic plot of the temperature difference against time. The fractional unaccomplished temperature on the Y-axis can be calculated by

$$Y = \ln\left(\frac{T - T_a}{T_1 - T_a}\right). \tag{4.1}$$

Published experimental diagrams allow a first approximation of temperature profiles and of chilling time at any chilling conditions. These nomographs can in turn be converted to simple prediction programs, as carried out by Drumm et al. (1992) on the nomographs of Bailey and Cox (1976) for chilling beef carcasses (Fig. 4.1). These models work well if there are no marked changes in thermal properties in the temperature range of interest.

4.2.2   Analytical Methods

Analytical methods assume a highly simplified mathematical model such that the basic differential equations can be solved analytically to yield simple formulae giving crude estimates of freezing or chilling time. These usually assume uniform temperature throughout the product, cooling at a constant temperature, and a uniform and constant heat transfer coefficient. Studies have been carried out in two areas: (1) the development of methods for shapes in which the heat transfer can be modeled as one-dimensional (infinite slabs, infinite cylinders, and spheres) and (2) the development of methods for shapes in which the modeling of the heat transfer must be two- or three-dimensional (Hossain et al., 1992). Most of the simplified equations are improvements of Plank's equation published in 1941 (López-Leiva and Hallström, 2003). The general equation is as follows:

$$t_F = \frac{rH_L}{T_F - T_\infty}\left(\frac{Pa}{h} + \frac{Ra^2}{K}\right). \tag{4.2}$$

The equation shows clearly that the freezing time (t_F) will increase with increasing density (ρ), latent heat of fusion (H_L), characteristic dimension (a), and difference between the initial freezing point (T_F) and the temperature of the freezing medium ($T\infty$), while an increase in the temperature gradient, the surface heat transfer coefficient (h), and the product thermal conductivity (K) will decrease the freezing time. The constants P and R are used to account for the influence of product shape, with $P = 1/2$, $R = 1/8$ for an infinite plate, $P = 1/4$, $R = 1/16$ for an infinite cylinder, and $P = 1/6$, $R = 1/24$ for a sphere.

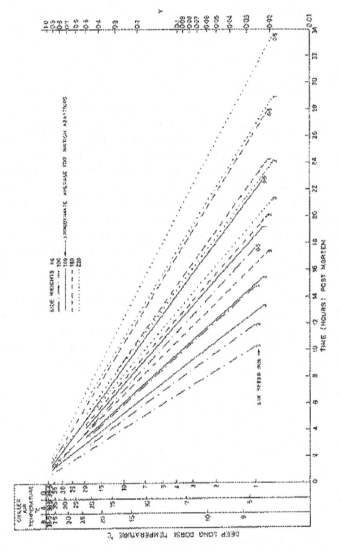

Fig. 4.1 Nomograph for calculating chilling time in the deep longissimus dorsi of a beef carcass

4.2.3   Numerical Methods

In numerical methods (finite differences, finite elements, finite volume), the product is divided into small control volumes or elements. The heat conduction equation is written for each of these control volumes/elements, giving a set of hundreds, thousands, or sometimes millions of equations. These equations are then solved together to calculate the change with time of the temperature in each location.

4.2.3.1   Finite Differences

Finite difference was the earliest numerical method. The product is represented by a regular grid and the partial differential equations for heat (or mass) flow are solved between the "nodes" (grid points) in time steps. The solution of a finite-difference problem on a computer is very fast, of the order of a few seconds or less, and the finite-difference equations can be quite easily set up. However, owing to mathematical difficulties, the use of these methods is in practice restricted to simple shapes (such as infinite and finite slab, infinite and finite cylinder, and sphere, the same as those used in analytical solutions). Their foremost advantage in relation to analytical modeling is that time variation in physical properties and boundary conditions can be easily introduced. Finite-difference models can be applied easily to a wide range of food products, particularly since many foods are contained in regular-shaped containers.

The following is an example of a finite-difference model that can be used for heat flow to and through an infinite slab of thickness X. This can be used for modeling chilling, freezing, or thawing operations. Since the slab is infinite, the heat flows only in one direction, so the system has one spatial dimension. The partial differential equation describing the process is the Fourier equation for heat conduction

$$\rho C_p \frac{\partial T}{\partial t} = \frac{\partial}{\partial x}\left(K\frac{\partial T}{\partial x}\right), \tag{4.3}$$

with an initial condition

$$T = T_{\text{start}}, \quad t = 0, 0 \le x \le X$$

and boundary conditions

$$K\frac{\partial T}{\partial x} = h_{\text{T}}\left(T_{\text{a}} - T_0\right), \quad t > 0, x = 0, \tag{4.4}$$

and

$$K\frac{\partial T}{\partial x} = h_{\text{B}}\left(T_{\text{a}} - T_0\right), \quad t > 0, x = X. \tag{4.5}$$

A numerical method based on that originally described by Dusinberre (1949) can be used. This approach was used by Bailey et al. (1974) and James and Bailey (1980) with layer equations being added by James and Swain (1982). With a numerical method, solutions are calculated only for discrete points (nodes) and for discrete times. The temperature at each node is taken as representative of a region that includes the node.

In the case of the slab, it is divided into a number of slices of thickness Δx, and we consider the nodes as shown in Fig. 4.2

The top and bottom surface nodes represent a region of thickness $\Delta x/2$, rather than Δx as for all the other nodes.

Considering an internal node and its region (Fig. 4.3), a heat balance in this region is given by equation 4.6.

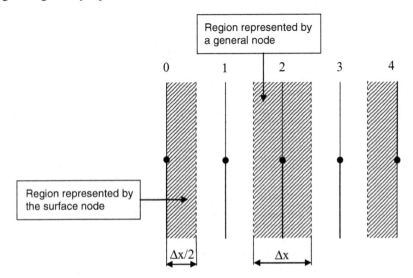

Fig. 4.2 Nodes through a slab

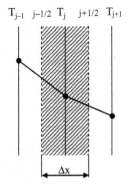

Fig. 4.3 Internal node

Rate of heat accumulation in the region (ϕ)= rate of heat transfer across (j + 1/2) boundary ($\phi_{j+1/2}$)– rate of heat transfer across (j + 1/2) boundary ($\phi_{j+1/2}$) or

$$\phi = \phi_{j+1/2} - \phi_{j-1/2}, \tag{4.6}$$

where ϕ is given by

$$\phi = V_j \rho C_{p_j} \frac{\partial T_j}{\partial t} = A\Delta x \rho C_{p_j} \frac{\partial T_j}{\partial t}. \tag{4.7}$$

Since the heat flow by conduction is defined by

$$KA\frac{\partial T}{\partial x}, \tag{4.8}$$

this gives

$$\phi_{j\pm1/2} = \left(KA\frac{\partial T}{\partial x}\right)_{j\pm1/2}. \tag{4.9}$$

Substituting Eqs. 4.7 and 4.9 in Eq. 4.6, one obtains the following equation:

$$A\Delta x \rho C_{p_j} \frac{\partial T_j}{\partial t} = \left(KA\frac{\partial T}{\partial x}\right)_{j+1/2} - \left(KA\frac{\partial T}{\partial x}\right)_{j-1/2}. \tag{4.10}$$

To create finite-difference approximations for the derivative and to produce an explicit model, a forward difference is used for $\partial T/\partial t$ and a central difference for $\partial T/\partial x$.

The forward difference is given by

$$\frac{\partial T_j}{\partial t} \approx \frac{T_j' - T_j}{\Delta t} \tag{4.11}$$

and the central difference is given by

$$\left.\frac{\partial T}{\partial x}\right|_{j+1/2} \approx \frac{T_{j+1} - T_j}{\Delta x}, \quad \left.\frac{\partial T}{\partial x}\right|_{j-1/2} \approx \frac{T_j - T_{j-1}}{\Delta x}. \tag{4.12}$$

Thermal conductivity values in j + 1/2, j – 1/2 boundaries can be approximated by an arithmetic mean value:

$$K_{j+1/2} = K_{j,j+1} = \frac{K(T_j) + K(T_{j+1})}{2}, \tag{4.13}$$

$$K_{j-1/2} = K_{j-1,j} = \frac{K(T_{j-1}) + K(T_j)}{2}. \tag{4.14}$$

Substituting Eqs. 4.11, 4.12, 4.13, and 4.14 in Eq. 4.10 gives

$$A\Delta x \rho C_{p_j} \frac{T'n_j - T_j}{\Delta t} = K_{j,j+1} A \frac{T_{j+1} - T_j}{\Delta x} - K_{j-1,j} A \frac{T_j - T_{j-}}{\Delta x} \tag{4.15}$$

and rearranging the equation for the future temperature in an internal node gives

$$T'_j = T_j + \frac{\Delta t}{\rho C_{p_j} \Delta x^2} \left[K_{j,j+1} \left(T_{j+1} - T_j \right) - K_{j-1,j} \left(T_j - T_{j-1} \right) \right]. \tag{4.16}$$

A similar procedure is used for the surface nodes. In this case, the heat is transferred from the ambient environment to the surface by convection and the surface exchanges heat with the adjacent node by conduction. This results in the following equation for the future temperature in the surface node:

$$T'_0 = T_0 + \frac{2\Delta t}{\rho C_{p_0} \Delta x} \left[K_{0,1} \frac{(T_1 - T_0)}{\Delta x} - h(T_0 - T_a) \right]. \tag{4.17}$$

In practice, foods such as bacon are not homogeneous but are made up of layers, e.g., fat and lean. These layers have very different thermal properties. The model can therefore be developed to handle multiple layers. Figure 4.4 shows the boundary between two such layers.

Δt must be the same for both materials, Δx can change, so region A is divided into slices of thickness Δx_A and region B into slices of thickness.Δx_B

K in the boundary surface is defined as

$$K_A = \frac{K(T_D)_{Mat\,A} + K(T_1)}{2} \tag{4.18}$$

and

$$K_B = \frac{K(T_D)_{Mat\,B} + K(T_2)}{2}. \tag{4.19}$$

The rate of heat accumulation (ϕ) inside the region is given by

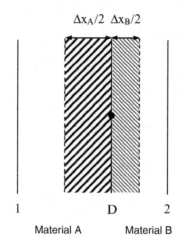

$\Delta x_A/2$ $\Delta x_B/2$

1   D   2

Material A   Material B

Fig. 4.4 Boundary between layers

$$\phi = A\left(\frac{\Delta x_A}{2}\rho_A CpA + \frac{\Delta x_B}{2}\rho_B C_{pB}\right)\frac{T'_D - T_D}{\Delta t}. \tag{4.20}$$

The rate of heat transfer across the boundary surface in material A is

$$\phi_A = K_A A\frac{T_D - T_1}{\Delta x_A}. \tag{4.21}$$

The rate of heat transfer across the boundary surface in material B is

$$\phi_B = K_B A\frac{T_2 - T_D}{\Delta x_B}. \tag{4.22}$$

Since $\phi = \phi_B - \phi_A$, when Eqs. 4.20, 4.21, and 4.22 are substituted, rearranged, and divided by A, one obtains the equation for the future temperature in the layer boundary:

$$T'_D = T_D + \frac{2\Delta t}{\rho_A C_{pA}\Delta x_A + \rho_B C_{pB}\Delta x_B}\left(K_B\frac{T_2 - T_D}{\Delta x_B} - K_A\frac{T_D - T_1}{\Delta x_A}\right). \tag{4.23}$$

A finite-difference scheme can become unstable and not converge to the correct solution for certain combinations of time steps and nodal spacing.

Rearranging Eq. 4.16 and making $K_{j,j+1} = K_{j-1,j}$ gives the following:

$$T'_j = \frac{\Delta t K}{\rho C_{p_j} \Delta x^2} \left[T_{j+1} + \left(\frac{\rho C p_j \Delta x^2}{\Delta t K} - 2 \right) T_j + T_{j-1} \right].$$

(4.24)

T_{j-1} and T_{j+1} are always multiplied by a positive factor for any value of Δx and Δt, but for T_j to be multiplied by a positive factor

$$\frac{\rho C_p \Delta x^2}{\Delta t K} - 2 \geq 0 \Rightarrow \Delta t \leq \frac{\rho C_p \Delta x^2}{2K}.$$

(4.25)

To ensure stability the program checks that Eq. 4.25 is met at the minimum value of C_p and the maximum value of K.

For the surface equation (Eq. 4.17), the factor that multiplies T_0 is

$$\frac{\rho C_p \Delta x}{2 \Delta t} - \frac{K}{\Delta x} - h \geq 0 \Rightarrow \Delta t \leq \frac{\Delta x^2 \rho C_p}{2(K + h \Delta x)},$$

(4.26)

again, evaluating at the minimum value of and maximum value of K and h.

From Eq. 4.23 for a boundary surface between different layers, obtains the following condition:

$$\frac{\rho_A C_{pA} \Delta x_A + \rho_B C_{pB} \Delta x_B}{2 \Delta t} - \frac{K_B}{\Delta x_B} - \frac{K_A}{\Delta x_A} \geq 0,$$

$$\Rightarrow \Delta t \leq \frac{\left(\rho_A C_{pA} \Delta x_A + \rho_B C_{pB} \Delta x_B \right) \Delta x_A \Delta x_B}{2 \Delta t}.$$

(4.27)

The condition in Eq. 4.27 is more stringent than that in Eq. 4.26, so it is enough to just check Eq. 4.26 for both surfaces and Eq. 4.27 for each boundary between layers.

The thermal properties of food are very temperature dependent. During operation, the model must obtain the relevant thermal properties as they change with temperature. These can be calculated from chemical properties using the equations of Miles et al. (1983).

The application of finite-difference-based methods to freezing and thawing presents some problems owing to the phase change that is reflected in the thermal properties as a sharp peak in the heat capacity curve and in a highly nonlinear thermal conductivity curve near the freezing point. The thermal conductivity problem can be dealt with quite simply just using the mean value. The heat capacity peak is a more serious problem. If too large a time step is used, the peak value of the specific heat may be missed. The temperature–enthalpy method proposed by Pham (1985) can be used to correct this problem. The equation is solved and it gives an approximated value of the future temperature; the heat gained by the node with this temperature gap is assumed to be correct, hence the future nodal enthalpy is

$$H' = \rho C_p (T' - T) + H.$$

(4.28)

The future temperature is then corrected according to this value of enthalpy in this way

$$(T')_{\text{Corrected}} = T(H').$$

(4.29)

This correction method reduces the heat balance percentage error (percentage difference between heat flows through boundaries and total heat gained of product) to less than 5%.

There are numerous sources of the equations for other shapes. Amongst these, infinite-cylinder equations can be found in James et al. (1977), sphere equations can be found in Bailey et al. (1974), and finite-cylinder equations can be found in Burfoot and James (1988). Considering an infinite cylinder of radius R or a sphere of radius R, we suppose that the heat flow is completely radial, so the system is still one-dimensional. The partial differential equation describing the process is the Fourier equation for heat conduction:

$$\rho C_p \frac{\partial T}{\partial t} = \frac{\partial}{\partial r}\left(K \frac{\partial T}{\partial r} \right) + \frac{ZK}{r}\frac{\partial T}{\partial r},$$

(4.30)

where Z is a constant factor depending on the shape

Mass transfer from the surface of the product can be added using the following equations based on those of Radford et al. (1976). The rate of mass transfer (kg s^{-1}) from or to the surface of a food by evaporation/condensation is described by

$$M = mA(P_s - P_a).$$

(4.31)

P_s and P_a are surface vapor pressure and air vapor pressure. Surface vapor pressure is calculated from relationship between saturated vapor pressure (P_{sat}) at the surface temperature and water activity (a_w). Air vapor pressure is calculated from the relationship between saturated vapor pressure at the air temperature and relative humidity (RH):

$$P_s = a_w P_{sat},$$

$$P_{sat} = f(T) : \text{evaluated at relevent surface temperature},$$

$$\text{RH} = \frac{P_a}{P_{sa}} \times 100\%,$$

$$P_a = \frac{\text{RH}\, P_{sat}}{100},$$

$$P_{sat} = f(T) : \text{evaluated at relevent air temperature}.$$

Surface saturated vapor pressure and air saturated vapor pressure are related to temperature. To calculate the rate of heat transfer due to evaporation/condensation, the rate of mass transfer (M) is multiplied by the latent heat (L).

Consequently, the rate of heat transfer (Q_e) from the surface of a food by evaporation/condensation is described by

$$Q_e = mLA(P_s - P_a).$$

(4.32)

When a warm product is transferred to a chiller or freezer some of the subsequent heat transfer will be due to radiation between the (hot) product surface and the (cold) environment. The affect of radiative heat transfer is related to the pattern of loading of the chiller/freezer. It is often incorporated into the overall surface heat transfer coefficient; however, it can be calculated relatively easily. The radiation heat transfer coefficient is related to the temperature difference between the radiation source and the environment, and will change as surface temperature changes. Most foods and radiating surfaces have similar values and an average value of 0.9 can be used for practical purposes.

The overall rate of heat transfer (Q) combines the radiation heat transfer coefficient (h_r) with the standard heat transfer coefficient (h_c):

$$Q = (h_c + h_r)(T_a - T_s) = h(T_a - T_s).$$

(4.33)

Thus, the effective heat transfer coefficient (h) is defined by

$$h = \frac{Q}{A(T_a - T_s)} = h_c + h_r.$$

(4.34)

To determine the rate of heat transfer (Q_r) by radiation from food, the following approximation may be applied:

$$Q_r = A\sigma\left(\varepsilon_s T_s^4 - \varepsilon_e T_e^4\right).$$

(4.35)

The radiation heat transfer coefficient (h_r) is defined by

$$h_r = \frac{\sigma\left(\varepsilon_s T_s^4 - \varepsilon_e T_e^4\right)}{(T_s - T_e)},$$

(4.36)

where ε_s and ε_e are the emissivity of the surface and the enclosure, respectively, A is the surface area of the food, σ is the Stefan–Boltzmann constant, T_s and T_e are the temperature of the surface and the enclosure, respectively.

As the temperature of the surface (and in some cases the enclosure) changes, the radiation heat transfer coefficient changes.

4.2.3.2 Finite Elements

Finite element is preferred to finite difference for modeling objects with complex, irregular shapes (Fig. 4.5). A grid is still used but it can be irregular, consisting of triangles, distorted rectangles, or volumes of various shapes. The grid can be in two or three dimensions. Finite element differs from finite difference in one major respect: finite difference assumes that thermal capacities are lumped at the nodes, Finite element assumes thermal capacity is distributed over the volume according to some weighting rule (Pham, 2001). The procedure is otherwise similar to that for finite difference and the equations obtained are assembled into a matrix equation for the nodal temperature vector T (Pham, 2001):

$$\frac{\mathbf{C}\mathrm{d}T}{\mathrm{d}t} = \mathbf{K}T^m + \mathbf{f}^m,$$

(4.37)

where $\mathbf{C}$ is the global capacitance matrix, $\mathbf{K}$ the thermal conductance (or stiffness) matrix, and $\mathbf{f}$ the thermal load vector. This matrix equation is then solved by a matrix-solving procedure.

Finite-element models take more time to solve than finite-difference models owing to the larger bandwith of the matrix equations that need to be solved. On a modern PC, a two-dimensional finite-element model may take several minutes to solve, while a three-dimensional finite-element model may take an hour or more. Also, it is more difficult to set up the model, usually requiring special graphical software. However, this is not due to the nature of the model itself, but to the complex

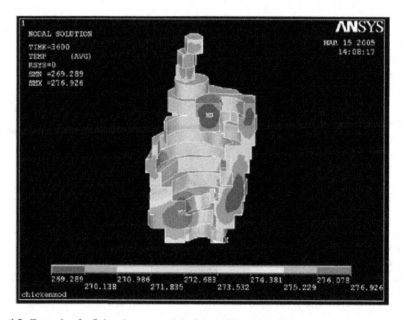

Fig. 4.5 Example of a finite-element model of the chilling of a chicken carcass

shape that such models are applied to. Finite-difference models are easier to set up because the real shape is usually approximated to a more regular shape.

4.2.3.3   Finite Volumes/Computational Fluid Dynamics

In recent years the method of finite volumes, which combines the flexibility of finite element with the conceptual simplicity of finite difference, has become widely popular (Pham, 2002). The object is divided into small control volumes of arbitrary shape (as in finite element), and the equations of conservation for heat, mass, and momentum are written for each control volume. The computational effort is similar for the finite-volume and finite-element methods. Finite volume is widely used in computational fluid dynamics (CFD) models (Pham, 2002; Wang and Sun, 2003).

In general, commercial CFD packages are utilized rather than creating models "from scratch." Xia and Sun (2002) reviewed many of the commercial CFD codes available. Few of these commercial packages have been developed with the needs of food refrigeration modelers in mind and much expertise and development time is required to utilize these programs.

One of the big uncertainties in calculating process time and heat load is the heat transfer coefficient at the surface of the product. Often the air is highly turbulent, and the shape is complex, giving rise to boundary layer variation and eddies, which make the heat transfer coefficient vary from place to place and become highly unpredictable. In other cases, such as cartoned product, the presence of air gaps with irregular shapes containing natural convection cells makes the effective heat transfer coefficient difficult to predict. CFD offers the promise of eliminating these uncertainties, by calculating the flow pattern and heat transfer at the product surface from first principles, using the basic equations of heat, mass, and momentum transfer (Pham, 2002).

For the case of laminar flow, CFD has indeed lived up to its promise. However, most industrial situations involve turbulent flow, for which CFD is still unable to entirely deliver the goods (Nicolai et al., 2001; Pham, 2002; Xia and Sun, 2002). This is because turbulent flow involves stochastic variations in the flow and temperature pattern, which must be averaged out, and during this averaging process, empirical equations and coefficients must be introduced and the equations lose their fundamentality. They must incorporate turbulence models that are still conceptually suspect and involve a large number of empirical coefficients. The empirical coefficients are specific to each flow situation and if they are used beyond these (i.e., in almost every new situation), accuracy is no longer guaranteed (Pham, 2002).

One other problem is that the thermophysical data of the materials of interest in the modeling of food refrigeration processes (vegetables, meat, composite foods, etc.) are often not available in the commercial CFD packages that are often aimed at the flow of air, water, etc. around and in solid objects.

Another obstacle against the widespread use of CFD models is the large amount of development time and computation effort involved. As an example, Pham (2002) writes that "it took about three months of an experienced user's time to build the finite volume of a beef side model, and simulating a 20-hour run took a week on a 300-MHz personal computer, with frequent intervention from the user to optimize relaxation factors etc. (and that's ignoring mass transfer)."

4.3   Surface Heat Transfer Coefficient

The rate of heat removal from the product depends on the surface area available for heat flow, the temperature difference between the surface and the medium, and the surface heat transfer coefficient. Each combination of product and cooling system can be characterized by a specific surface heat transfer coefficient. The value of the coefficient depends on the shape and surface roughness of the foodstuff, but not its composition, and to a much greater degree on the thermophysical properties and velocity of the cooling medium. Typical values range from 5 W m^{-2} K^{-1} for slow-moving air to 500 W m^{-2} K^{-1} for agitated water. Becker and Fricke (2004) analyzed 777 cooling curves from 295 food items to obtain heat transfer coefficients for industrial processes. In industrial applications the type of packaging applied has a significant effect on the value of the heat transfer coefficient.

Heat transfer coefficients may be determined experimentally, using the transient temperature method (Creed and James, 1985) or a steady-state sensor (Harris et al., 2004). Alternatively the heat transfer coefficient may be calculated using a correlation between a set of dimensionless numbers: Nusselt number, Prandtl number, Reynolds number, and Grashof number. A wide range of empirical equations have been collected by Krokida et al. (2002) and Zogzas et al. (2002). Often the heat transfer coefficient is determined by fitting predicted temperatures to experimental data through trial and error. Increasingly CFD is also being considered (Wang and Sun, 2003).

In air thawing, h is not constant and is a function of RH (James and Bailey, 1982). In the initial stages, water vapor condenses onto the frozen surface, immediately changing to ice. This is followed by a stage where vapor condenses in the form of water until the surface temperature is above the dew point of the air and all condensation ceases. The varying rate of condensation produces substantial changes in the value of h during the thawing process.

The effect of errors in the assumed heat transfer coefficient on the estimation of freezing time has been analyzed by Fricke and Becker (2006).

4.4   Thermal Properties

In the cooling of foodstuffs the thermal properties of the products play a very important role. For homogeneous materials there are few problems in applying theoretical models for heat transfer within solids as well as between a solid surface and a liquid. However, nonhomogeneous materials, to which most food products belong, often present problems and the literature covering this field is much more sparse. Secondly, the physical properties may vary remarkably, not only with origin, but also owing to previous treatments. Moreover, variations in the properties also take place during cooling. For many foods there are few published data; thus, theoretical models that can calculate food properties over a temperature range from the chemical composition of the foods are attractive (Jowitt et al., 1983). Saad and Scott (1996) have made suggestions for how to estimate thermal properties of basic food products during freezing, and Fricke and Becker (2006) recently quantitatively evaluated selected thermophysical property models.

Accurate knowledge of the initial freezing point is particularly important for freezing, thawing, and tempering models. This can be determined experimentally or using many of the, largely empirical, published models (James et al., 2005). Experimentally it can be more accurately determined from the thawing data curve than from the freezing data curve.

4.5   Chilling

Heat and mass transfer models for predicting chilling processes have been recently reviewed by Cleland (1990), Pham (2001), and Wang and Sun (2003). Specific recent reviews of meat chilling modeling have been provided by Daudin and van Gerwen (1996) and Kuitche et al. (1996a).

Time–temperature charts for infinite slabs, infinite cylinders, and spheres can be found in Singh and Heldman (1993) and ASHRAE Fundamentals (2001). These are derived from numerical methods of unsteady-state heat conduction and can be used to gain a quick answer for chilling times by assuming simple shapes and constant thermal properties with temperature. To avoid the inaccuracies of manually reading off a chart, the equations used to generate the information in the charts can be put onto a computer.

Empirically based diagrams have been published by a number of researchers that allow a first approximation of temperature profiles and of chilling time at any chilling condition. Some of these data have been supplemented by numerically modeled data. Many of these are for meat carcasses: beef (James and Bailey, 1990), pork (Brown and James, 1992), lamb and mutton (Earle and Fleming, 1967), and goat (Gigiel and Creed, 1987). Similar systems have also been used for the cooling of fish in ice (Jain et al., 2005), the cooling of trays of ready meals in air (Evans et al., 1996), and immersion systems (Ketteringham and James, 1999).

In the last two cases, finite-difference methods were used to extrapolate the experimental data to a wider range of products and chilling conditions. James and Swain (1982) used a finite-difference model based on an infinite cylinder to study the influence of surface fat layers on the chilling time of beef carcasses. The model was later extended (James and Schofield, 1998) to produce a simple predictive program (BeefChill) to study the chilling of beef sides. Finite-difference techniques have also been used to model the immersion chilling of foods (Zorrilla and Rubiolo, 2005a, b).

Finite-element modeling has been used to study heat and mass transfer in the chilling of chicken in a continuous system (Landfeld and Houska, 2006). The mathematical modeling employed the codes Bertix (TNO-MEP, Holland) and Food Product Modeler (Mirinz, Food Technology and Research, New Zealand). A three-dimensional explicit finite-difference mathematical model was developed to investigate air-impingement cooling of food in the shape of finite slabs (Erdogdu et al., 2005).

There has been relatively little use of CFD to model chilling. It has been used to model cooling times of cooked meat joints in an air blast-chilling process (Hu and Sun, 2001). Trujillo and Pham (2003) used an evolutionary algorithm to adjust unknown parameters during the beef cooling process, while Mirade et al. (2002) have looked at its use in designing large food chillers.

Several models have been developed specifically for vacuum cooling of liquid foods. Vacuum cooling is different from conventional air and immersion cooling methods as it involves the internal generation of the cooling source due to water evaporation under vacuum pressure; thus, any modeling involves coupled heat transfer with inner heat generation and mass transfer with inner vapor generation (Wang and Sun, 2003). One of the earliest models was that of Burfoot et al. (1989), which was further developed by Houska et al. (1996). A further model was developed by Dostal and Petera (2004). Wang and Sun (2002a, b) have modeled the vacuum cooling of cooked meats.

4.6 Freezing

A great many methods of predicting freezing times for foodstuffs have been proposed. Since there are interesting mathematical problems, owing to changing thermal properties, phase changes, etc., involved with freezing, substantially more work has been published on modeling the freezing of foodstuffs than on modeling chilling, even though in practice far more food is chilled than is frozen. This represents the difference between academic interest and industrial interest. Heat and mass transfer models for predicting freezing processes have recently been reviewed by Cleland (1990), Hung (1990), Pham (2001), and Delgado and Sun (2001).

Plank (1913) first presented as a formula for calculating the freezing time for a block of ice, then subsequently presented a similar calculation for food products (Plank, 1941). Many have used this classic equation as a starting point and López-Leiva and Hallström (2003) have reviewed its interpretation and modification. There are a number of useful publications where many similar models are compared, for example, LeBlanc et al. (1990) and López-Leiva and Hallström (2003). An interesting combination of numerical modeling and a simple analytical method has been used to predict weight loss during the freezing and storage of unwrapped food (Campanone et al., 2005a) and to predict freezing time (Campanone et al., 2005b).

A simple, one-dimensional, infinite-slab, finite-difference model was used successfully to predict the freezing of meat blocks (James and Bailey, 1979). Creed and James (1984) used a similar model based on an infinite cylinder to predict freezing times in mutton carcasses. Huan et al. (2003) used a finite-element model to look at quick freezing of foods, while Moraga and Barraza (2003) used finite-volume techniques to look at the freezing of plate-shaped foods. Impingement freezing technology, where high-velocity air jets are directed against the food surface to break the insulating boundary layer that surrounds the product, has been modeled by Salvadori and Mascheroni (2002).

4.7   Thawing

Thawing has received much less attention in the literature than either chilling or freezing (Lind, 1991). In commercial practice there are relatively few controlled thawing systems. Thawing is often considered as simply the reverse of the freezing process. However, inherent in thawing is a major problem that does not occur in the freezing operation. The majority of the bacteria that cause spoilage or food poisoning are found on the surfaces of meat. During the freezing operation, surface temperatures are reduced rapidly and bacterial multiplication is severely limited, with bacteria becoming completely dormant below −10°C. In the thawing operation these same surface areas are the first to rise in temperature and bacterial multiplication can recommence. On large objects subjected to long uncontrolled thawing cycles, surface spoilage can occur before the central regions have fully thawed. Also the thermal diffusivity of ice is approximately 9 times larger than that of water (Bakal and Hayakawa, 1973).

Finite-difference models have been successfully used to predict the thawing of meat blocks (James and Bailey, 1980; Creed and James, 1981) and catering packs (Brown et al., 2006), while Hoke et al. (2002) has used models to look at the sensitivity of the thawing process to changes in environmental parameters.

There has been some progress in modeling of microwave thawing in recent years with models by Chamchong and Data (1999a, b) and Basak and Ayappa (2002) based on the thawing of tylose.

The process of tempering has received even less coverage. Studies by Brown (Brown, 1997; Brown and James, 2006) used a similar finite-difference model to that used by James and Bailey (1980) and Creed and James (1981) to determine the effects of air temperature and velocity on tempering times for beef blocks of various compositions. A slightly more sophisticated finite-difference model was used by James et al. (2003) to model two-stage tempering systems for bacon joints.

4.8   Distribution

Much of the effort on modeling refrigeration processes has concentrated on the modeling of refrigeration processes that change the temperature of the food, such as chilling, freezing, and thawing. The purpose of refrigerated distribution is to maintain the temperature of the food and appears to have attracted less attention from modelers. It is particularly important that the food is at the correct temperature before distribution since the refrigeration systems used in most transport containers, storage rooms, and retail display cabinets are not designed to extract heat from the load, but to maintain the temperature of the load.

CFD and other techniques have been used to model the performance of chilled and frozen storage rooms. Studies have been carried out on airflow in cold stores (Said et al., 1995; Mariotti et al., 1995; Hoang et al., 2000) and through cold store doorways (Foster et al., 2002). Tanner et al. (2002a–c) have developed models for heat and mass transfer in packaged horticultural products in storage.

In the large containers used for long-distance transportation, food temperatures can be kept within ±0.5°C of the set point. With this degree of temperature control, transportation times of 8–14 weeks (for vacuum-packed meats stored at −1.5°C) can be achieved and the products can still retain sufficient chilled storage life for retail display. However, there are substantial difficulties in maintaining the temperature of refrigerated foods transported in small-refrigerated vehicles that conduct multidrop deliveries to retail stores and caterers. During any one delivery run, the refrigerated product can be subjected to as many as 50 door openings, where there is heat ingress directly from outside and from personnel entering to select and remove product. The design of the refrigeration system has to allow for extensive differences in load distribution, dependent on different delivery rounds, days of the week, and the removal of product during a delivery run (Tso et al., 2002). In the UK the predictive program CoolVan has been produced to aid the design and operation of small refrigerated delivery vehicles (Gigiel, 1998).

There are stages in transportation where food is not in a refrigerated environment, that is, it is in loading bays, in supermarkets before loading into retail displays, in domestic transportation from shop to home, etc. Some food transportation models have looked at temperature rises in pallet loads of chilled or frozen food during distribution. They often specifically look at the times during loading, unloading, and temporary storage when the pallets may be in a warm environment that may cause the food temperature to rise (Bennahmias et al., 1997; Stubbs et al., 2004). Under European quick-frozen food regulations, frozen products must be distributed at −18°C or lower with brief upward fluctuation of no more than 3°C allowed within the distribution. Models have shown that in the case of open loading bays the initial temperature of the frozen product needs to be below −25°C to keep it within the regulations.

The final stages of the refrigerated food chain, retail display and domestic refrigeration, has also attracted the attention of modelers. Again CFD (Fig. 4.6) and other numerical techniques have been used to model air movement in retail display cabinets (Baleo et al., 1995; Van Ort and Van Gerwen, 1995; Lan et al., 1996; Stribling et al., 1997; Navaz et al., 2002; Cortella, 2002).

Laguerre and Flick (2004) have developed a model to look at heat exchange within domestic refrigerators and the time required to cool any warm food that is placed in them, while Anderson et al. (2004) have looked at thawing and freezing of meat products in domestic appliances.

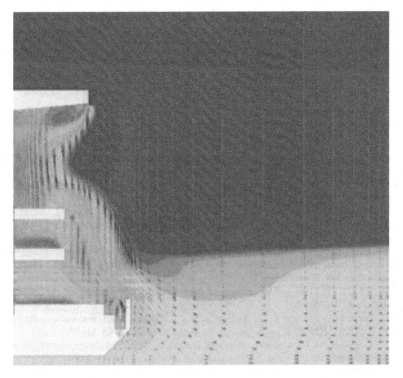

Fig. 4.6 Computational fluid dynamics model of the loss of cold air from a refrigerated display cabinet

4.9   The Future

Cleland (1990) has stated that there are four factors that limit the accuracy of freezing and chilling time prediction. In general, the same factors limit the accuracy of predicting any food refrigeration process.

The four factors are:

- Imprecise knowledge of the conditions surrounding the product in the system
- Imprecise thermal and diffusional data for the product
- Extrapolating a method beyond its range of applicability
- Shortcomings in the prediction method

He states that in every case the observed difference between a measured freezing or chilling time and that predicted is the net effect resulting from the interaction and accumulation of the four factors.

Although the accuracy of refrigeration models has increased over time, the complexity has also increased, with greater resources required, such as time, computing facilities, and a good working knowledge of mathematics. These models, however, have not always been put into practice by those working in the industry owing to the lack of conceptual understanding and people's perception of them. For this to happen there needs to be a greater emphasis on simple-to-use, flexible, and fast programs.

References

Anderson B. A., Sun S., Erdogdu F., Singh R. P. (2004): Thawing and freezing of selected meat products in household refrigerators. *International Journal of Refrigeration*, 27:1, 63–72

ASHRAE (2001): Fundamentals. American Society of Heating, Refrigeration and Air-Conditioning Engineers, Inc., Atlanta, GA

Bakal A., Hayakawa K. I. (1973): Heat transfer during freezing and thawing of foods. In: Advances in Food Research (eds. C. O.Chichester, E. M.Mrak and G. F.Stewart), Academic, New York, Vol. 20, pp. 218–256

Baleo J. N., Guyonnaud L., SolliecC. (1995): Numerical simulations of air flow distribution in a refrigerated display case air curtain. *Proceedings of the 19th International congress of Refrigeration*, Vol. 2, pp. 681–687

Bailey C., Cox R. P. (1976): The chilling of beef carcasses. *Proceeding of the Institute of Refrigeration*, 72, 76–90

Bailey C., James S. J., Kitchell A. G., Hudson W. R. (1974): Air-, water- and vacuum-thawing of frozen pork legs. *Journal of the Science of Food and Agriculture*, 25, 81–97

Basak T., Ayappa K. G. (2002): Role of length scales on microwave thawing dynamics in 2D cylinders. *International Journal of Heat and Mass Transfer*, 45, 4543–4559

Becker B. R., Fricke B. A. (2004): Heat transfer coefficients for forced-air cooling and freezing of selected foods. *International Journal of Refrigeration*, 27:5, 540–551

Bennahmias R., Gaboriau R., MourehJ.(1997): The insulating cover, a particular logistic means for thermo-sensitive foodstuffs. *International Journal of Refrigeration*, 20:5, 359–366

Brown T. (1997): Tempering times for boxed boneless beef. In: Engineering and Food at ICEF 7 Supplement, pp. 9–12

Brown T., James S. J. (1992): Process design data for pork chilling. *International Journal of Refrigeration*, 15:5, 281–289

Brown T., James S. J. (2006): The effect of air temperature, velocity and block composition on the tempering time of frozen boneless beef blocks. *Meat Science*, 73:4, 545–552

Brown T., Evans J., James C., JamesS. J., Swain M. J. (2006): Thawing of cook-freeze catering packs. *Journal of Food Engineering*, 74:1, 70–77

BurfootD., JamesS. J.(1988): The effect of spatial variations of heat transfer coefficient on meat processing times. *Journal of Food Engineering*, 7, 41–61

Burfoot D., Hayden R., Badran R. (1989): Simulation of a pressure cook/water and vacuum cooled processing system. In: Process Engineering in the Food Industry: Developments and Opportunities (eds. R. W. Field, J. Howell), Elsevier Applied Science, London, pp. 27–41

Campanone L. A., Salvadori V. O., Mascheroni R. H. (2005a): Food freezing with simultaneous surface dehydration: approximate prediction of weight loss during freezing and storage. *International Journal of Heat and Mass Transfer*, 48:6, 1195–1204

Campanone L. A., Salvadori V. O., Mascheroni R. H. (2005b): Food freezing with simultaneous surface dehydration: approximate prediction of freezing time. *International Journal of Heat and Mass Transfer*, 48:6, 1205–1213

Chamchong M., Data A. K. (1999a): Thawing of foods in a microwave oven: I. *Effect of power levels and power cycling. Journal of Microwave Power and Electromagnetic Energy*, 34, 9–21

Chamchong M., Data A. K. (1999b): Thawing of foods in a microwave oven: II. *Effect of load geometry and dielectric properties. Journal of Microwave Power and Electromagnetic Energy*, 34, 22–32

Cleland A. C. (1990): Food refrigeration processes – Analysis, design and simulation. Elsevier Applied Science, London, England, 284 p

Cortella G. (2002): CFD aided retail cabinets design. *Computers and Electronics in Agriculture*, 34, 43–66

Creed P. G., James S. J. (1981): Predicting thawing times of frozen boneless beef blocks. *Proceedings of the Institute of Refrigeration*, 77, 355–358

Creed P. G., James S. J. (1984): The prediction of freezing and thawing times of mutton carcasses. Proceedings of the 30th European Meeting of Meat Research Workers, Bristol, 2.5, pp. 59–60

Creed P. G., James S. J. (1985): Heat transfer during the freezing of liver in a plate freezer. *Journal of Food Science*, 50, 285–288, 294

Daudin J. D., van Gerwen R. J. M. (1996): Methods to Assess Chilling Kinetics in Industrial Chillers. New developments in Meat Refrigeration, IIR Antony, France, pp. 1.7–1.15

Delgado A. E., Sun D. W. (2001): Heat and mass transfer models for predicting freezing processes – a review. *Journal of Food Engineering*, 47, 157–174

Dostal M., Petera K. (2004): Vacuum cooling of liquids: mathematical model. *Journal of Food Engineering*, 61:4, 533–539

Drumm B. M., Joseph R. L., Mckenna B. M. (1992): Line chilling of beef 1: the prediction of temperatures. *Journal of Food Engineering*, 16, 251–265

Dusinberre G. M. (1949): Numerical Analysis of Heat Flow (1st edn). McGraw-Hill, New York

Earle R. L., Fleming K. A. (1967): Cooling and freezing of lamb and mutton carcasses: -1- cooling and freezing rates in legs. *Food Technology*, 21, 79–84

Erdogdu F., Sarkar A., Singh R. P. (2005): Mathematical modelling of air-impingement cooling of finite slab shaped objects and effect of spatial variation of heat transfer coefficient. *Journal of Food Engineering*, 71:3, 287–294

Evans J., Russell S., James S. (1996): Chilling of recipe dish meals to meet cook-chill guidelines. *International Journal of Refrigeration*, 19, 79–86

Foster A. M., Barrett R., James S. J., Swain M. J. (2002): Measurement and prediction of air movement through doorways in refrigerated rooms. *International Journal of Refrigeration*, 25, 1102–1109

Fricke B. A., Becker B. R. (2006): Sensitivity of freezing time estimation methods to heat transfer coefficient error. *Applied Thermal Engineering*, 26:4, 350–362

Gigiel A. (1998): Modelling the thermal response of foods in refrigerated transport, Meeting of IIR Commission D1,D2/3 Cambridge, UK, International Institute of Refrigeration, Paris

Gigiel A. J., Creed P. G. (1987) : Effect of air speed and carcass weight on the cooling rates and weight losses from goat carcasses. *International Journal of Refrigeration*, 10, 305–306

Harris M. B., Carson J. K., Willix J., Lovatt S. J. (2004): Local surface heat transfer coefficients on a model lamb carcass. *Journal of Food Engineering*, 61:3, 421–429

Hoang M. L., Verboven P., De Baerdemaeker J., Nicolai B. M. (2000): Analysis of the air flow in a cold store by means of computational fluid dynamics. *International Journal of Refrigeration*, 23, 127–140

Hoke K., Houska M., Kyhos K., Landfeld A. (2002): Use of a computer program for parameter sensitivity studies during thawing of foods. *Journal of Food Engineering*, 52:3, 219–225

Hossain M. M., Cleland D. J., Cleland A. C. (1992): Prediction of freezing and thawing times for foods of regular multi-dimensional shape by using an analytically derived geometric factor. *International Journal of Refrigeration*, 15, 227–234

Houska M., Podloucky S., Zitny R., Gree R., Sestak J., Dostal M., Burfoot D. (1996): Mathematical model of the vacuum cooling of liquids. *Journal of Food Engineering*, 29:3–4, 339–348

Hu Z., Sun D. W. (2001): Effect of fluctuation in inlet airflow temperature on CFD simulation of air-blast chilling process. *Journal of Food Engineering*, 48, 311–316

Huan Z., He S., Ma Y. (2003): Numerical simulation and analysis for quick-frozen food processing. *Journal of Food Engineering*, 60:3, 267–273

Hung Y. C. (1990): Prediction of cooling and freezing times. *Food Technology*, 44:5, 137–144, 146, 148,153

Jain D., IlyasS. M., Pathare P., Prasad S., Singh H.(2005): Development of mathematical model for cooling the fish with ice. *Journal of Food Engineering*, 71:3, 324–329

James S. J., Bailey C. (1979): The determination of the freezing time of boxed meat blocks. *Proceedings of the Institute of Refrigeration*, 75, 1–8

James S. J., Bailey C. (1980): Air and vacuum thawing of unwrapped boneless meat blocks. *Proceedings of The Institute of Refrigeration*, 76, 44–51

James S. J., Bailey C. (1982): Changes in the surface heat transfer coefficient during meat thawing. Proceedings of the 28th European Meeting of Meat Research Workers, Madrid, 1, 16 March, 160–163

James S. J., Bailey C. (1990): Chilling of beef carcasses. Chilled foods – The state of the art, Elsevier Applied Science, London, England, Chapter 8, pp. 159–182

James S. J., Schofield I. (1998): Developments in the prediction of meat carcass chilling. In: Advances in the Refrigeration System, Food Technologies and Cold Chain. Proceedings of meeting of IIF-IIR Commissions B2 and C2, with D1 and D2/3, Sofia, Bulgaria

James S. J., Swain M. J. (1982): The effect of surface fat layers on the chilling time of meat. *Proceedings of the 16th International Congress of Refrigeration, Paris*, Vol. 2, 473–478

James S. J., Creed P. G., Roberts T. A. (1977): Air thawing of beef quarters. *Journal of the Science of Food and Agriculture*, 28, 1109–1119

James C., Palpacelli S., James S. (2003): Optimisation of two-stage bacon tempering using mathematical modelling. In: Predictive Modelling in Foods – Conference Proceedings (eds. J. F. M. Van Impe, A. H. Geeraerd, I. Leguérinel, P. Mafart), Katholieke Universiteit Leuven/BioTeC, Belgium, ISBN 90-5682-400-7, pp. 277–279

James C., Lejay I., Tortosa N., Aizpurua X., James S. J. (2005): The effect of salt concentration on the freezing point of meat simulants. *International Journal of Refrigeration*, 28, 933–939

Jowitt R., Escher F., Hallstrom B., Meffert H. F. T., Spiess W. E. L., VosG.(1983): Physical Properties of Foods. Applied Science, London

Ketteringham L., James S. J. (1999): Immersion chilling of trays of cooked products. *Journal of food Engineering*, 40, 256–267

Krokida M. K., ZogzasN. P., MaroulisZ. B.(2002): Heat transfer coefficient in food processing: compilation of literature data. *International Journal of Food Properties*, 5:2, 435–450

Kuitche A., Daudin J. D., Letang G. (1996a): Modelling of temperature and weight loss kinetics during meat chilling for time-variable conditions using an analytical-based method.1. The model and its sensitivity to certain parameters. *Journal of Food Engineering*, 28:1, 55–84

Laguerre O., Flick D. (2004): Heat transfer by natural convection in domestic refrigerators. *Journal of Food Engineering*, 62:1, 79–88

Lan T. H., Gotham D. H. T., Collins M. W. (1996): A numerical simulation of the air flow and heat transfer in a refrigerated food display cabinet. Second European Thermal Sciences and 14th UIT National Heat Transfer Conference, pp. 1139–1146

Landfeld A., Houska M. (2006): Prediction of heat and mass transfer during passage of the chicken through the chilling tunnel. *Journal of Food Engineering*, 72:1, 108–112

LeBlanc D. I., Kok R., Timbers G. E. (1990): Freezing of a parallelepiped food product. *Part 2, Comparison of experimental and calculated results. International Journal of Refrigeration*, 13, 379–392

Lind I.(1991): Mathematical modelling of the thawing process. *Journal of Food Engineering*, 14, 1–23

López-Leiva M., Hallström B. (2003): The original Plank equation and its use in the development of food freezing rate predictions. *Journal of Food Engineering*, 58, 267–275

Mariotti M., Rech G., Romagnoni P. (1995): Numerical study of air distribution in a refrigerated room. Proceedings of the 19th International Conference of Refrigeration, August 20–25, The Hague, The Netherlands, pp. 98–105

Miles C. A., van Beek G., Veerkamp C. H. (1983): Calculation of thermophysical properties of foods. In: Physical Properties of Foods (eds. R. Jowitt, F. Escher, B. Hallstrom, H. F. Th. Meffert, W. E. L. Spiess, G. Vos), Applied Science, London, Chapter 16, pp. 269–312

Mirade P. S., Kondjoyan A., Daudin J. D. (2002): Three-dimensional CFD calculations for designing large food chillers. *Computers and Electronics in Agriculture*, 34:1–3, 67–88

Moraga N. O., Barraza H. G. (2003): Predicting heat conduction during solidification of a food inside a freezer due to natural convection. *Journal of Food Engineering*, 56:1, 17–26

Navaz H. K., Faramarzi R., Gharib M., Dabiri D., Modaress D.(2002): The application of advanced methods in analyzing the performance of the air curtain in a refrigerated display case. *Journal of Fluids Engineering – Trans ASME*, 124:3, 756–764

Nicolai B. M., Verboven P., Scheerlinck N., Hoang M. L., Haddish N.(2001): Modelling of cooling and freezing operations. Rapid cooling of food, Meeting of IIR Commission C2, Bristol, UK, Section 3, pp. 211–216

Pham Q. T. (1985): A fast, unconditionally stable finite-difference method for heat conduction with phase change. *International Journal of Heat Mass Transfer*, 28, 2079–2084

Pham Q. T. (2001): Modelling thermal processes: cooling and freezing. In: Food Process Modelling (eds. L. M. M. Tijskens, M. L. A. T. M. Hertog, B. M. Nicolaï), Woodhead Publishing Limited, Cambridge, Chapter 15

Pham Q. T. (2002): Calculation of processing time and heat load during food refrigeration. EcoLibrium: The Official Journal of AIRAH, July, 22–28

PlankR. (1913): Die Gefrierdauer von Eisblöcken (Freezing times of ice blocks). *Zeitschrift für die gesamte Kälte-Industrie*, XX:6, 109–114

Plank R.(1941): Beiträge zur Berechnung und Bewertung der Gefriergeschwindigkeit von Lebensmittel (Calculation and validation of freezing velocities in foods). *Beihefte zur Zeitschrift für die gesamte, Kälte-Industrie*, 3:10, 22

Radford R. D., Herbert L. S., Lovett D. A. (1976): Chilling of meat – a mathematical model for heat and mass transfer. Towards an ideal refrigerated food chain, Meeting of IIR Commissions C2, D1, D2, D3, and E1, Melbourne, Australia, pp. 323–330

Saad Z., Scott E. P. (1996): Estimation of temperature dependent thermal properties of basic food solutions during freezing Journal of Food Engineering, 28, 1–19

Said M. N. A., Shaw C. Y., Zhang J. S., Christianson L. (1995): Computation of room air distribution. *ASHRAE Transactions*, 101, 1065–1077

Salvadori V. O., Mascheroni R. H. (2002): Analysis of impingement freezers performance. *Journal of Food Engineering*, 54:2, 133–140

Singh R. P., Heldman D. R. (1993): Introduction to Food Engineering (2nd edn). Clarendon, Oxford

Stribling D., Tassou S. A., Marriot D. (1997): A two dimensional CFD model of a refrigerated display case. *ASHRAE Transactions: Research*, 103:1, 88–95

StubbsD. M., PulkoS. H., WilkinsonA. J.(2004): Wrapping strategies for temperature control of chilled foodstuffs during transport. *Transactions of Instrumental Measurement and Control*, 26:1, 69–80

Tanner D. J., Cleland A. C., Opara L. U. (2002a): A generalised mathematical modelling methodology for the design of horticultural food packages exposed to refrigerated conditions: Part 2. Heat transfer modelling and testing. *International Journal of Refrigeration*, 25:1, 43–53

Tanner D. J., Cleland A. C., Opara L. U., Robertson T. R. (2002b): A generalised mathematical modelling methodology for design of horticultural food packages exposed to refrigerated conditions: Part 1. Formulation. *International Journal of Refrigeration-Revue*, 25:1, 33–42

Tanner D. J., Cleland A. C., Robertson T. R. (2002c): A generalised mathematical modelling methodology for design of horticultural food packages exposed to refrigerated conditions: Part 3. Mass transfer modelling and testing. *International Journal of Refrigeration*, 25:1, 54–65

Trujillo F. J., Pham Q. T. (2003): Modelling the chilling of the leg, loin and shoulder of beef carcasses using an evolutionary method. *International Journal of Refrigeration*, 26:2, 224–231

Tso C. P., Yu S. C. M., Poh H. J., Jolly P. G. (2002): Experimental study on the heat and mass transfer characteristics in a refrigerated truck. International Journal of Refrigeration, 25, 340–350

Van Ort H., Van Gerwen R. J. M. (1995): Air flow optimisation in refrigerated cabinets. Proceedings of 19th International Congress of Refrigeration, pp. 446–453

Wang L., Sun D. W. (2002a): Modelling vacuum cooling process of cooked meat--part 1: Analysis of vacuum cooling system. *International Journal of Refrigeration*, 25:7, 854–861

Wang L., Sun D. W. (2002b): Modelling vacuum cooling process of cooked meat – part 2: Mass and heat transfer of cooked meat under vacuum pressure. *International Journal of Refrigeration*, 25:7, 862–871

Wang L., Sun D. W. (2003): Recent developments in numerical modelling of heating and cooling processes in the food industry – a review. *Trends in Food Science and Technology*, 14, 408–423

Xia B., Sun D. W.(2002): Applications of computational fluid dynamics (CFD) in the food industry: a review. *Computers and Electronics in Agriculture*, 34, 5–24

Zogzas N. P., Krokida M. K., Michailidis P. A., Maroulis Z. B. (2002): Literature data of heat transfer coe fficients in food processing. *International Journal of Food Properties*, 5:2, 391–417

Zorrilla S. E., Rubiolo A. C. (2005a): Mathematical modelling for immersion chilling and freezing of foods: Part i: Model development. *Journal of Food Engineering*, 66:3, 329–338

Zorrilla S. E., Rubiolo A. C. (2005b): Mathematical modelling for immersion chilling and freezing of foods: Part ii: Model solution. *Journal of Food Engineering*, 66:3, 339–351

Notation

P	Density (kg m^{-3})
A	Area (m^2)
Cp	Specific heat (J kg^{-1}°C^{-1})
H	Specific enthalpy (J kg^{-1})
H	Surface heat transfer coefficient (W m^{-2}°C^{-1})
K	Thermal conductivity (W m^{-1}°C^{-1})
L	Latent heat of boiling
M	Mass transfer coefficient (kg m^{-2} s^{-1} Pa^{-1})
P_a	Air vapor pressure
P_s	Surface vapor pressure
R	Distance, radius (m)
T	Temperature (°C)
T	Time (s)
t_F	Freezing time (s)
T_a	Ambient (environment) temperature (°C)
Ti	Current temperature of node i for the slab, infinite cylinder, and sphere model (°C)
Ti	Future temperature of node i for the slab, infinite cylinder, and sphere model (°C)
Δx	Space increment (m)
Δr	Radial increment (m)
Δt	Time increment (s)

Some other specific notation is explained in the relative paragraph.

Chapter 5
Modeling of Heating During Food Processing

Ivanka Zheleva and Veselka Kamburova

5.1 Introduction

Heat transfer processes are important for almost all aspects of food preparation and play a key role in determining food safety. Whether it is cooking, baking, boiling, frying, grilling, blanching, drying, sterilizing, or freezing, heat transfer is part of the processing of almost every food. Heat transfer is a dynamic process in which thermal energy is transferred from one body with higher temperature to another body with lower temperature. Temperature difference between the source of heat and the receiver of heat is the driving force in heat transfer.

To conduct effective process modeling there is a need for consistent thermo-physical property information.

5.2 Thermal Properties of Foods

The thermal properties of food products determine their ability to transfer and store heat. The thermal properties are *specific heat*, Cp (J/kg/K), *thermal conductivity*, k (W/m/K), and *thermal diffusivity*, α (m²/s).

5.2.1 Specific Heat

Specific heat is a measure of the amount of energy required by a unit mass to raise its temperature by a unit degree. So, specific heat is the quantity of heat that is gained or lost by a unit mass of product to accomplish a unit change in temperature, without a change in state. It can be calculated as follows:

Iv. Zheleva (✉)
Rousse University, Studentska Str. 8, Rousse 7017 Bulgaria

R. Costa, K. Kristbergsson (eds.), *Predictive Modeling and Risk Assessment,*
DOI: 10.1007/978-1-387-68776-6, © Springer Science+Business Media, LLC 2009

$$C_p = \frac{Q}{m\Delta T},$$

(5.1)

where Q is the heat gained or lost (kJ), m is the mass (kg), ΔT is the temperature change in the material (K), and Cp is the specific heat (kJ/kg/K).

Specific heat is an essential part of the thermal analyses of food processing or of the equipment used in heating or cooling of food. A number of models express specific heat as a function of water content, as water is a major component of many foods. Siebel (1892) proposed that the specific heat of food materials such as eggs, meats, fruits, and vegetables can be taken as equal to the sum of the specific heat of water and solid matter. One of the earliest models to calculate specific heat was proposed by Siebel (1892) and Charm (1978). Siebel's model is described by the following equation:

$$C_p = 0.837 + 3.349W,$$

(5.2)

where W is the water content expressed as a fraction of the total base.

The influence of product components was expressed in an empirical equation proposed by Charm (1978) as

$$C_p = 2.093X_f + 1.256X_s + 4.187W,$$

(5.3)

where X_f is the mass fraction of fat and X_s is the mass fraction of nonfat solids.

Other equations of form similar to that of Eq. 5.2 have been summarized by Sweat (1986).

Choi and Okos (1986) suggested a more generalized equation for specific heat which takes into account the composition of food:

$$C_p = 4.180W + 1.711X_p + 1.928X_f + 1.547X_c + 0.908X_a,$$

(5.4)

where X is mass or weight fraction of each component and the subscripts denote the following components: p – protein, f – fat, c – carbohydrate, and a – ash.

Although specific heat varies with temperature, for ranges near room temperature, these changes are relatively minor. They are usually neglected in engineering calculations. Sweat (1986) gave several equations for specific heat which include temperature dependency.

5.2.2 Thermal Conductivity

The thermal conductivity of a food material is an important property used in calculations involving the rate of heat transfer. Thermal conductivity k is the rate of heat transfer q through a unit cross-sectional area A when a unit temperature difference (T_1-T_2) is maintained over a unit distance L:

$$k = \frac{qL}{A(T_1 - T_2)}.$$

$$(5.5)$$

Equation 5.5, which implies steady-state heat transfer conditions, has been used to design experiments for measuring thermal conductivity of food. Transient techniques are also used for more rapid determination of thermal conductivity. These experimental methods have been reviewed by Choi and Okos (1986). Experimental data on thermal conductivities measured for various food groups have been expressed by mathematical relationships. These models are useful in estimating thermal conductivity of food materials. Riedel's (1949) model predicts thermal conductivity of fruit juices, sugar solutions, and milk:

$$k = \left(1326.58 + 1.0412T - 0.00337T^2\right)(0.46 + 0.54W)1.73 \times 10^{-3}.$$

$$(5.6)$$

Here W is the mass fraction of water. This formula gives an error in the temperature interval between 0 and 180°C of approximately 1%.

Most high-moisture foods have thermal conductivity values close to that of water. On the other hand, the thermal conductivity of dried, porous food is influenced by the presence of air, which has low thermal conductivity. Empirical equations are useful in process calculations where the temperature may be changing. For fruits and vegetables with water content greater than 60% the following equation was proposed by Sweat and Haugh (1974):

$$k = 0.148 + 0.493W.$$

$$(5.7)$$

Another empirical equation was developed by Sweat (1986) for solid and liquid food:

$$k = 0.25X_c + 0.155X_p + 0.16X_f + 0.135X_a + 0.58W.$$

$$(5.8)$$

Choi and Okos (1986) suggested the following model for liquid food:

$$k = \sum k_i X_i^V,$$

$$(5.9)$$

where the estimated volume fraction, $X_i^V = \dfrac{X_i^W / \rho_i}{\sum (X_i^W / \rho_i)}$, where X_i^W is the mass fraction of its components and ρi its density, and ki is its thermal ductivity. Thermal conductivities of anisotropic materials vary with the direction of heat transfer. For example, the thermal conductivity along the meat fibers is different from that across the fibers.

5.2.3   Thermal Diffusivity

Thermal diffusivity, α, is a ratio involving thermal conductivity, density, and specific heat and can be expressed as

$$\alpha = \frac{k}{\rho C_p}. \tag{5.10}$$

Here α is the thermal diffusivity (m²/s) and ρ is the density (kg/m³).

If thermal conductivity, density, and specific heat are known, thermal diffusivity can be calculated. Thermal diffusivity is strongly influenced by the water content as shown by the following models:

• Dickerson's (1969) model,

$$\alpha = 0.088 \times 10^{-6} + (\alpha_w - 0.088 \times 10^{-6})W \tag{5.11}$$

• Marten's (1980) model,

$$\alpha = 0.057363W + 0.00028(T + 273) \times 10^{-6} \tag{5.12}$$

Here α_w is the thermal diffusivity of water. Choi and Okos (1986) suggested the following model for liquid food:

$$\alpha = \sum \alpha_i X_i^V, \tag{5.13}$$

where αi is thermal conductivity of the components of the food.

Thermal properties depend on the composition of the food, its temperature, and its water content. These dependences usually cannot be neglected and must be taken into account.

5.3   Modes of Heat Transfer

There are three modes of heat transfer: *conduction*, *convection*, and *radiation*. Any energy exchange between bodies occurs through one of these modes or combinations of two or all three modes.

5.3.1   Conduction

In conduction, the molecular energy is directly exchanged from the hotter to the cooler regions, the molecules with greater energy communicating some of this energy to neighboring molecules with less energy. The effectiveness by which heat is transferred through a material is measured by the thermal conductivity, k (W/m/K). The rate of heat transfer by conduction q_{cond} (W) is given by

$$q_{cond} = -kA\frac{dT}{dx},$$

(5.14)

where A is the area of cross section of the heat flow path (m²) and dT/dx is the temperature gradient, that is the rate of change of temperature per unit length of path (K/m). Equation 5.14 is known as the Fourier equation for heat conduction. In heat transfer, a positive q means that heat is flowing into the body, and a negative q represents heat leaving the body. A minus sign appears in Eq. 5.14 because heat flows from a hotter to a colder body and that is in the direction of the temperature gradient.

Fourier's law may be solved for a rectangular, cylindrical, or spherical coordinate system, depending on the geometrical shape of the object being studied.

Conduction heat transfer in a slab of material is given by

$$q_{cond} = -kA\frac{dT}{dx} = kA\frac{T_1 - T_2}{x}.$$

(5.15)

Here T_1 and T_2 ($T_1 > T_2$) are the temperatures of the faces of the slab and x is the thickness of the slab (see Fig. 5.1).

Conduction heat transfer in a three-layered wall where the thermal conductivities of the three layers are k_1, k_2, and k_3 and the thicknesses of each layer are x_1, x_2, and x_3, respectively, is given by (see Fig. 5.2)

$$q_{cond} = A\frac{T_1 - T_2}{\frac{x_1}{k_1} + \frac{x_2}{k_2} + \frac{x_3}{k_3}}.$$

(5.16)

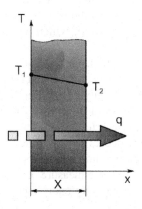

Fig. 5.1 A plane wall

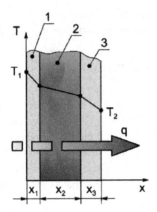

Fig. 5.2 A composite wall

Conduction heat transfer in a hollow pipe is give by

$$q_{cond} = \frac{2\pi Lk(T_1 - T_2)}{\ln\left(\dfrac{r_2}{r_1}\right)},$$

(5.17)

where L is the length of the pipe, r_1 is the inside radius of the pipe, and r_2 is the outside radius.

5.3.2 Convection

Convection is the transfer of heat by the movement of groups of molecules in a fluid. The groups of molecules may be moved by either density changes or forced motion of the fluid. In a typical convective heat transfer a hot surface heats the surrounding fluid, which is then carried away by fluid movement. The warm fluid is replaced by cooler fluid, which can draw more heat away from the surface.

There are two types of convection: natural convection and forced convection. Heat transfer by *natural convection* occurs when a fluid is in contact with a surface hotter or colder than itself. The density of the fluid decreases as it becomes hotter and the density increases as the fluid is cooled. The difference in density causes movement in the fluid that has been heated or cooled and causes the heat transfer to continue. *Forced convection* occurs when a fluid is forced to pass a solid body and heat is transferred between the fluid and the body.

The convection coefficient, h, is the measure of how effectively a fluid transfers heat by convection. It is determined by many factors, such as the fluid density, viscosity, and velocity, and the geometrical shape of the object undergoing heating or cooling. The convection heat transfer coefficient, h, is measured in watts per square meter per kelvin. The rate of heat transfer from a surface by convection is given by Newton's law:

$$q_{cond} = -hA(T_s - T_\infty).$$ (5.18)

Here A is the surface area of the object, T_s is the surface temperature, and T_∞ is the ambient or fluid temperature.

5.3.3   Radiation

Radiation heat transfer is the transfer of heat energy by electromagnetic waves, which transfer heat from one body to another, in the way that electromagnetic light waves transfer light energy. Radiation heat transfer occurs when the emitted radiation strikes another body and is absorbed.

The basic formula for radiation heat transfer is the Stefan–Boltzmann law:

$$q_{rad} = A\sigma T^4.$$ (5.19)

Here T is the absolute temperature in kelvins, σ is the Stefan–Boltzmann constant, $\sigma = 5.73 \times 10^{-8}$ W/m^2/K^4, and A is the area in square meters. This formula gives the radiation emitted by a *black body*, which gives the maximum amount of emitted radiation possible at its particular temperature. Real surfaces are usually *gray bodies* and for them the Stefan–Boltzmann law can be written as

$$q_{rad} = \varepsilon A\sigma T^4,$$ (5.20)

where ε is the material property called emissivity. ε has a value between 0 and 1, and is a measure of how efficiently a surface emits radiation. The radiation heat transferred between two surfaces depends upon their temperatures, the geometrical arrangement, and their emissivities. For two parallel surfaces, the net heat transferred from the hotter surface with temperature T_1 and emissivity ε_1 to the cooler surface with temperature T_2 and emissivity ε_2 is given by

$$q_{rad} = AC\sigma \left(T_1^4 - T_2^4\right),$$ (5.21)

where

$$C = \frac{1}{\dfrac{1}{\varepsilon_1} + \dfrac{1}{\varepsilon_2} - 1}.$$

5.4   Mass Transfer

In many practically important cases mass transfer plays a key role in food processing. If there are differences in concentrations of constituents throughout a solution or object, there will be a tendency for movement of material to produce a uniform concentration. Such movement may occur in gas, liquid, or solid solutions, Movement resulting from random molecular motion is called diffusion. Mass transfer occurs during various food processing operations, such as humidification and dehumidification, dehydration, distillation, and absorption. The driving force for mass diffusion is the concentration difference. The basic relationship is called Fick's law and can be written as follows (Singh and Heldman, 2001):

$$N_A = -D \frac{dC_A}{dx}, \tag{5.22}$$

where N_A is the mass flux of species A (kg/s/m^2), C_A is the concentration measured in mass per volume, x is the distance in the direction of diffusion, and D is the diffusion coefficient, or diffusivity. Most of the heat and mass transfer modeling in terms of cooking has been concentrated on specific food and biological products.

5.5   Methods of Applying Heat to Food

Foods are heated by indirect or direct methods. In indirect heating, heat is applied to the food through heat exchangers, the products of combustion being isolated from the food. In direct systems, the heat energy is passed directly into the food.

The methods of applying heat to food may be classified as follows:

- *Indirect heating* by vapors or gases such as steam or air; liquids such as water and organic heat-exchange liquids; electricity, in resistance heating systems.
- *Direct heating* using gas, oil, and solid fuels; using infrared energy; using electricity, by dielectric or microwave methods.

5.5.1   Indirect Heating Methods

These systems comprise four components: a combustion chamber where the fuel is burned and combustion products disposed of; a heat exchanger where the heat of combustion is taken up by a heat transfer fluid; a transfer system where the heated transfer fluid is passed to the heat user; and a heat exchanger where the transfer fluid exchanges its heat with the food. Indirect heating is illustrated in Fig. 5.3.

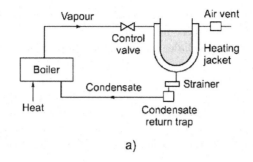

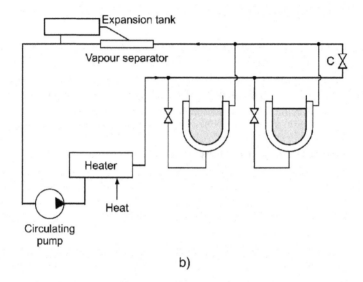

Fig. 5.3 Simple fluid heating systems: **a** vapor system; **b** liquid system

5.5.1.1 Indirect Heating by Vapors or Gases

Steam and air are commonly used heat transfer media. Steam may be used for heating in its saturated or its superheated form; it may be also used to operate electric generators and vacuum ejectors. No other material possesses these unique properties. That is why steam is an essential requirement in most food plants.

When saturated steam is applied as a heat transfer medium it has a high latent heat and a high thermal conductivity, which are advantageous. But saturated steam has some disadvantageous properties, such a high vapor pressure and a low critical point. Steam is very suitable for food processes because it is nontoxic, fireproof and explosion-proof, and odorless; it is produced from a cheap and abundant raw material – water. In normal practice, saturated steam is used for food processing up to a temperature in the region of 200°C. Above this, the cost of the necessary high-pressure equipment starts to become unduly high.

Superheated steam finds little heating application in the food industry for sterilization.

Air is a poor heat transfer fluid since it has low specific heat and thermal conductivity. Nevertheless, air is used for the heating of canned food in the Ekelund cooker, for baking, for drying, and in fluidized-bed cooking. In all these cases heat transfer is by forced convection. Air is, of course, nontoxic and noncontaminating although it can bring about deterioration in foods which are sensitive to oxidation.

5.5.1.2   Indirect Heating of Food by Liquids

Liquids such as water, mineral oils, chlorinated hydrocarbons, and fused salts are used for general process heating. High-temperature water is a most useful medium at temperatures up to 200°C when advantage may be taken of its high specific heat and thermal conductivity. The other liquids find application in higher-temperature processing since they have the advantage of low vapor pressures.

5.5.1.3   Indirect Heating of Food by Electrical Resistance Heating

Electrical resistance heating is the generation of heat by the flow of current through a resistor. The resistors may be attached to the walls of the process vessels or immersed in the material to be heated (immersion heating). The heating elements are made from spirally wound, nickel–chromium wires. These elements work at temperature up to 800°C, so heat transfer into the food is primarily conductive. Resistance-heated baking ovens are common; resistor banks are located within the oven, heat being conveyed to the food by a combination of conduction, convection, and radiation.

5.5.2   Direct Heating Methods

There are some risks using the direct heating of food by solid, gaseous, and liquid fuels but numerous direct-fired baking ovens, malt kilns, and driers are encountered in the food industry.

Direct heating, by means of electrode boilers, is used in special steam-boiling applications.

5.5.2.1   Infrared Heating of Food

Infrared heating occurs by means of banks of radiant heaters located in a tunnel, where food is conveyed, or in a oven, where food is baked. Radiant heaters are of two types: medium-temperature heaters and high-temperature heaters.

Medium-temperature heaters work in the temperature range 500–1,000°C. High-temperature heaters operate at a temperature of about 2,500°C. Although some of the heat is transferred convectively, the major part of the energy is radiated in the infrared with a range of wavelengths between 0.75 and 350 μm.

Radiant energy is transformed into heat only by absorption, a process described by Lambert's law (Isachenko et al., 1981):

$$\frac{q}{q_0} = \exp(-\alpha x), \tag{5.23}$$

where q is the amount of radiation transferred to depth x in the material, q_0 is the incident intensity of radiation, and α is the absorption coefficient of the heated material. Practical systems involve selective radiators so that absorption varies from zero ($\alpha \rightarrow 0$) to complete absorption ($\alpha \rightarrow \infty$) for different wavelengths. Wavelengths up to about 50 μm are of practical importance in food heating. Water and aqueous systems absorb best at wavelengths around 1 μm.

When absorption of infrared energy occurs, it is characterized by low penetration, and produces rapid surface cooking of the food. This results in rapid sealing and browning of the other layers. Penetration of the center of the food piece is mainly by conduction, which is often a slow process. It is important to ensure that an adequate final center temperature is obtained.

Radiant heating has many applications in food processing, such as grilling, toasting, baking, specialized dehydratation procedures, and melting of fats.

5.5.2.2   Dielectric Heating of Food

Dielectric heating and microwave heating use high-frequency energy to avoid interference in radar, television, and radio transmissions. The permitted frequencies below 300 MHz, which are called radio frequencies, are used in dielectric heating and those above 300 MHz are called microwaves and are used in microwave heating. The phenomena of dielectric heating and microwave heating are essentially similar. The differences are only due to the different frequencies used, which determine the extent of energy penetration and the type of equipment employed.

Dielectric heating is defined as heating in an electrically insulating material by the losses in it when subjected to an alternating electric field. The material to be heated constitutes a dielectric sandwiched between capacitor plates connected to a capacitive, high-frequency, alternating generator. Heating is brought about by molecular friction due to the rapid orientation of the electric dipoles under the influence of the high-frequency alternation of the applied field.

Dielectric heating equipment generally comprises a low-loss belt, which conveys the food at a controlled rate between the plates of the capacitor. The top plate may be raised or lowered to control the heat generation in the product, and power input is also variable. Heaters capable of dissipating 160 kW of energy in the food with an efficiency of about 50% are usual, although units generating many hundreds of kilowatts have been installed.

It must be pointed that the rate of dielectric heating is very fast compared with that of conventional heating; local overheating is minimized. Those two factors reduce heat damage to the food. Dielectric heating is clean, continuous in operation, and well suited to automatic control. It is used to thaw frozen eggs, meat, fruit juices, and fish, to melt fats, chocolate, and butter, to bake biscuits, to heat peanuts, and to dry sugar cubes and crisp bread.

5.5.2.3 Microwave Heating of Food

Microwaves are regarded at electromagnetic radiation having frequencies in the range 300–300,000 MHz. Microwave radiation belongs to the group of nonionizing forms of radiations, since it does not have sufficient energy required for the ionization process. The quantum energy in microwaves is responsible for creation of heat as the microwave oscillates $2,450 \times 10^6$ times per second and the dipole molecules align to the electric field of the microwave at the same rate. The alternating electric field stimulates the oscillation of the dipoles of the molecules (e.g., water) in the food. The heat is generated owing to the molecular friction between dipole molecules.

Microwaves are generated by a magnetron. A magnetron is a circular symmetric tubelike diode that consists of a cathode as the central axis of the tube and an anode around the circumference. The magnetron contains a space called resonant cavities. Resonant cavities act as tuned circuits and generate electric fields (Fig. 5.4).

These cavities also determine the output frequency of the microwave. The magnetron has an antenna connected to an anode and it extends into the resonant cavities. The antenna is used for transmitting the microwave from the magnetron to the waveguide. The magnetic field is created by magnets, which surround the magnetron (Fig. 5.5) (Saltiel and Datta, 1998; Knutson et al., 1987).

When power is supplied, an electron-emitting material at the cathode becomes excited and emits electrons into the vacuum space between the cathode and the anode. The energy of the electrons is caught in the fields. The excess microwave

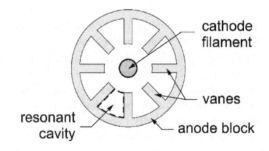

Fig. 5.4 Anode block

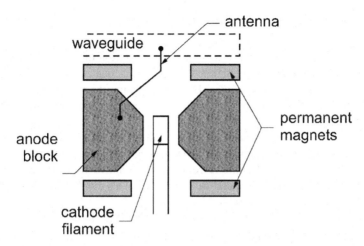

Fig. 5.5 Side view of the magnetrons

energy travels as waves and is extracted by the antenna. The antenna transmits the oscillating waves to the waveguide, where they travel into the oven cavity. The waveguide is a hollow metal tube. The metallic walls of the waveguide are nearly perfect electric conductors and the microwaves propagate with low transmission losses. As the waves enter the cavity, they are dispersed by a stirrer. This action minimizes hot and cold spots in the oven cavity. Normally, the magnetron operates with efficiency around 60–65% (Saltiel and Datta, 1998). Inside the microwave oven, waves can be reflected off the oven sides and floor, can be transmitted through containers and lids, and can be absorbed by food.

Microwaves are used to heat precooked, frozen food and, in conjunction with the browning effect of infrared heaters, for cooking in canteens and hospitals, where speed is important. Industrially, microwaves are used for precooking chicken, for apple juice evaporation, and in potato chips finishing. There are some investigational works for using microwaves in the pasteurization of fruit juices, in reducing mold counts in bread, cakes, and jam, in bread baking, and in accelerated freeze-drying.

5.6 Heat Methods of Food Safety and Food Preservation

Blanching, pasteurization, and sterilization are heating process with the objective of ensuring the preservation of food; the last two also accomplish the objective of safety. Other processes such as baking, roasting, and frying also accomplish these objectives, but their main purpose is to transform food materials for consumption (Anon, 1984).

5.6.1 Blanching

Blanching is an important heat process in the preparation of vegetables and fruits for canning, freezing, or dehydratation. Blanching inactivates enzymes or destroys enzyme substrates such as peroxides. During blanching the food is heated rapidly to a predetermined temperature, it remains at this temperature for a predetermined time, and then the food cools.

Two methods of blanching are used: immersion blanching using hot water, and steam blanching. Immersion blanching involved passing the food at a controlled rate through a perforated drum, and rotation in a tank of water thermostatically controlled to the blanching temperature (75–95°C). This method leads to high loss of soluble nutrients in some food. (Selman, 1987).

Steam blanchers utilize saturated steam at atmospheric or at low pressure (150 kPa). The food is conveyed through the steam chamber on a mesh belt, and the residence time is controlled by the conveyer speed. The blanched product is discharged through an outlet lock to a washer and cooler. Steam blanching gives lower blanching losses than immersion blanching. Steam blanchers are easier to use for sterilization than water blanchers.

Microwave blanching has been applied to fruits and vegetables packaged in film bags (Anon., 1981) and would appear to offer some advantages, such as microbiological cleanliness and low losses of nutrients.

Blanching problems arise in ensuring uniform heat treatment, and in controlling blanching losses and the effluent disposal difficulties caused by these. These problems are alleviated using individual quick blanching (Lazar et al., 1971). This is a modified three-stage steam blanching in which pieces of food are rapidly heated in a thin layer followed by holding as a deep bed, where equilibration takes place, after which the food is cooled by chilled air. This permits a shorter retention time and reduces effluent volumes and strength.

5.6.2 Sterilization

The purpose of sterilization is to destroy all microorganisms present in the food material to prevent its spoilage and to ensure it is safe for consumption. Microorganisms are destroyed by heat, but the amount of heating required to kill different organisms varies. Also, many bacteria can exist in two forms, the vegetative or growing form and the spore or dormant form. The spores are much harder to destroy by heat treatment than are the vegetative forms.

The time and the temperature required for the sterilization of food are influenced by several factors, including the type of microorganisms found on the food, the size of the container, the acidity or pH of the food, and the method of heating (Encyclopedia Britannica, 2006).

The thermal processes of canning are generally designed to destroy the spores of the bacterium *Clostridium botulinum*. The ratio of the initial to the final number of surviving organisms becomes the criterion that determines adequate treatment. For *C. botulinum*, this number is $10^{12}:1$. The mean time is called the thermal death time for the corresponding temperature. The standard reference temperature is generally selected as 121.1°C, and the selected time (in minutes) required to sterilize any selected organism at 121°C is referred to as the *F* value of that organism. For *C. botulinum*, the *F* value is about 2.8 min. If the total value so found is below 2.8 min, the sterilization is not sufficient; if it is above 2.8 min, the heat treatment is more drastic than it needs to be.

The other factor that must be determined, so that the equivalent killing powers at temperatures different from 121°C can be evaluated, is the dependence of thermal death time on temperature. Experimentally it has been found that if the logarithm of the thermal death time is plotted against the temperature, a straight-line relationship is obtained. The very sharp decrease of thermal death times with higher temperatures means that holding times at the lower temperatures contribute little to the sterilization. Very long times at temperatures below 90°C would be needed to make any appreciable differences to *F*, and in fact it can often be the holding time at the highest temperature which virtually determines the *F* value of the whole process.

5.6.3 Pasteurization

Pasteurization is a heat treatment which is sufficient to inactivate pathogenic microorganisms present in foods. This heating method of treating food is less drastic than sterilization. It aims to inactivate pathogenic organisms such as bacteria, viruses, protozoa, molds, and yeasts, but not harm the flavor or quality of the food. The process was named after its inventor, French scientist Louis Pasteur. The first pasteurization test was completed by Pasteur and Claude Bernard in 1862 (Encyclopedia Britannica, 2006).

Nowadays milk, wine, beer, and fruit juices are all routinely pasteurized. Pasteurization is also used with cheese and egg products. The most common application is pasteurization of liquid milk. There are two methods of pasteurization of milk which are commonly used. In the conventional method food is heated at least at 63°C and kept at that temperature for at least 30 min. In the other method the temperature is higher (71°C) and food is kept at that temperature for at least 15 s.

Unlike sterilization, pasteurization is not intended to kill all microorganisms in the food. Instead, pasteurization aims to achieve a "log reduction" in the number of viable organisms, reducing their number so they are unlikely to cause disease (assuming the pasteurized product is refrigerated and consumed before its expiration date). Pasteurization inactivates most viable vegetative forms of microorganisms but not heat-resistant spores. Pasteurization also may be considered in relation to enzymes present in the food, which can be inactivated by heat. A combination of temperature and time must be used that is sufficient to inactivate the particular species of bacteria or the enzyme under consideration. The temperature and time requirements of the pasteurization process are influenced by the pH of the food. When the pH is below 4.5, spoilage microorganisms and enzymes are the main target of pasteurization. For example, the pasteurization process for fruit juices is aimed at inactivating certain enzymes such as pectinesterase and polygalacturonase. The typical processing conditions for the pasteurization of fruit juices include heating to 77°C and holding for 1 min, followed by rapid cooling to 7°C. In addition to inactivating enzymes, these conditions destroy any yeasts or molds that may lead to spoilage. Equivalent conditions capable of reducing spoilage microorganisms involve heating to 65°C and holding for 30 min or heating to 88°C and holding for 15 s. When the pH of a food is greater than 4.5, the heat treatment must be severe enough to destroy pathogenic bacteria. For example milk pasteurization aims at inactivation of *Mycobacterium tuberculosis*, *Coxiela burnetti*, and *Brucella abortus*. For this process a time/temperature curve is constructed and can be sued to determine the necessary holding time and temperature.

The processes for sterilization and pasteurization illustrate very well the application of heat transfer as a unit operation in food processing. The temperatures and times required are determined and then the heat transfer equipment is designed using the equations developed for heat transfer operations.

5.7   Modeling of Heating of Food

Mathematical modeling of cooking processes plays an important role in designing and optimizing the cooking process.

As mentioned earlier, whenever there is a temperature difference in a medium or between media, heat transfer occurs (Incropera and De Witt, 1990). The theoretical and empirical relationships utilized in the design of heat processes assume knowledge of the thermal properties of the material, as discussed in Sect. 5.2. Heat transfer between a solid and its surroundings can take place by conduction, convection, and radiation. In some cases, all three forms of heat transfer operate simultaneously.

In this section the partial differential equations which describe heat and mass transfer in food processing are presented and discussed.

5.7.1   Heat Transfer Equation

In food process engineering, the transfer of heat occurs very often in the unsteady state, when materials are warming or cooling. Unsteady-state (or transient) heat transfer is that phase of the heating or cooling process when the temperature changes as a function of both location and time. The governing equation describing unsteady-state heat transfer in solid bodies in a 3D rectangular coordinate system is

$$\frac{\partial T}{\partial t} = \alpha \left(\frac{\partial^2 T}{\partial x^2} + \frac{\partial^2 T}{\partial y^2} + \frac{\partial^2 T}{\partial z^2} \right). \tag{5.24}$$

For solution of Eq. 5.24 written in terms of partial differentials in three dimensions it is desirable to determine the relative importance to heat transfer. For this purpose, a dimensionless ratio, called the Biot number (Bi), is useful:

$$Bi = \frac{hL}{k} = \frac{L}{1/h}. \tag{5.25}$$

The Biot number is the ratio of internal resistance and external resistance; hence the Biot number provides a measure of the temperature change in the solid relative to the temperature difference between the surface and the fluid.

In Eq. 5.25, h is the convective heat transfer coefficient, k is the thermal conductivity, and L is a characteristic dimension.

When the Biot number is small (smaller than 0.1), the product has very small internal resistance to heat transfer, that is, the thermal conductivity of the object is high. A high Biot number (greater than 100) implies that external resistance to heat transfer is large. For a Biot number between 0.1 and 100 there is a finite resistance to heat transfer both internally and at the surface of the object undergoing heating or cooling (Isachenko et al., 1981).

The study of heat flow under these conditions is complicated. There are some cases in which this equation can be simplified and handled by elementary methods, and charts have been prepared which can be used to obtain numerical solutions under some conditions of practical importance. A simple case of unsteady-state heat transfer arises from the heating or cooling of solid bodies made from good thermal conductors.

For example, the governing equation (Eq. 5.24) may be solved analytically for a long slab, a long cylinder, or a sphere. For a long slab Eq. 5.24 in one dimension is written as follows:

$$\frac{\partial T}{\partial t} = \alpha \frac{\partial^2 T}{\partial x^2}. \tag{5.26}$$

The boundary conditions are

$$-\kappa \frac{\partial T(s,t)}{\partial x} = \alpha T(s,t) \, for \, x = s$$

and

$$\frac{\partial T(0,t)}{\partial x} = 0 \, for \, x = 0. \tag{5.27}$$

where λ is the thermal conductivity (W/m/K),

The dimensionless temperature is

$$\theta = \frac{T - T_m}{T_0 - T_m},$$ (5.28)

where T_m is the medium temperature and T_0 is the initial temperature.

Equation 5.21 with appropriate initial and boundary conditions (Eq. 5.22) has an analytical solution. The following solution is obtained for a long slab in rectangular coordinates:

$$\theta = \sum_{n=1}^{\infty} \frac{2}{\mu_n} (-1)^{n+1} \cos \frac{\mu_n r}{R} \exp\left(-\mu_n^2 \frac{\alpha t}{R^2}\right),$$ (5.29)

where R is the half-thickness of a slab, r is the variable distance from the center axis, t is time, and μ_n is defined as

$$\mu_n = (2n-1)\frac{\pi}{2}.$$ (5.30)

Equation 5.24 may be presented in dimensionless form as follows:

$$\theta = \sum_{n=1}^{\infty} \frac{2}{\mu_n} (-1)^{n+1} \cos \mu_n X \exp(-\mu_n^2 Fo),$$ (5.31)

where $X = \dfrac{r}{R}$ is a dimensionless coordinate and $Fo = \dfrac{\alpha t}{R^2}$ is the Fourier number.

In Eq. 5.26 $\cos(\mu_n X)$ is a function of the Biot number; therefore the dimensionless temperature θ is a function of the Biot number and the Fourier number:

$$\theta = F(Bi, Fo).$$ (5.32)

In Fig. 5.6 are given the temperature distributions in a long slab during cooling for different values of the Biot number as follows: Fig. 5.6a – $Bi \to \infty$; Fig. 5.6b – $Bi \to 0$; Fig. 5.6c – $0.2 < Bi < 40$. All cases are presented for the same values of the Fourier number and $Fo_1 < Fo_2 < Fo_3 < Fo_4$.

Similar to the present case, the governing equation (Eq. 5.26) is solved analytically to obtain the solution for temperature in a finite slab:

$$\theta = \sum_{n=1}^{\infty} \frac{2 \sin \mu_n}{\mu_n + \sin \mu_n \cos \mu_n} \cos \mu_n X \exp(-\mu_n^2 Fo) = F(Bi, Fo).$$ (5.33)

The solutions of Eq. 5.24 for long and finite slabs, long and finite cylinders, and spheres have been reduced to charts that are much easier to use.

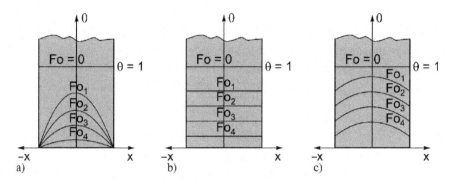

Fig. 5.6 Temperature distribution in a long slab with $Fo_1 < Fo_2 < Fo_3 < Fo_4$: **a** $Bi \to \infty$; **b** $Bi \to 0$; **c** $0.1 < Bi < 100$

5.7.2 Simultaneous Heat and Mass Transfer

Most food substances have complex structures, which can be represented as a hygroscopic capillary porous medium. Moisture migration in the food can be attributed to several different mechanisms, such as molecular diffusion, capillarity, bulk transport, effusion, and surface diffusion. All of these mechanisms may be important under a particular set of conditions, depending on the moisture content and the type of food product. Simultaneous heat and moisture exchange is used intensively in food processing operations such as drying, smoking, baking, frying, and cooking. Mathematical models are useful tools in the design and analysis of these processes. With utilization of the different mechanisms, several mathematical models have been developed to describe the simultaneous transfer of heat and moisture which predict the distribution of moisture content and temperature in food (Balaban, 1989; Luikov, 1975).

Because of the difficulty in using the models, different simplifying assumptions have been proposed. The usual assumption is that heat processing in the food occurs under isothermal conditions, which obviates the coupling between heat and moisture transfer (Balaban, 1989).

The first simultaneous heat and moisture transfer model was developed by Luikov (1975). He described the moisture transfer in porous materials such as dough products by a number of diffusion mechanisms, and by capillary and hydrodynamic flow. This was represented mathematically by a system of partial differential equations. Assuming constant thermophysical properties, Luikov provided an analytical solution to the model equations. Without the assumption of constant properties, the Luikov model becomes nonlinear and obtaining an exact solution is not feasible. The assumption of constant thermophysical properties may lead to erroneous predictions of temperature and moisture content.

The mathematical model consists of a couple of partial differential equations for simultaneous mass and heat transfer. The model is based on the assumption that the material is structured as a porous slab. The model equations are

$$\frac{\partial T}{\partial \tau} = \frac{\partial}{\partial x}\left(\alpha \frac{\partial T}{\partial x}\right) + \varepsilon \frac{\rho}{c_\rho} \frac{\partial u}{\partial \tau} + \frac{Q_v}{c_\rho} \qquad (5.34)$$

and

$$\frac{\partial u}{\partial \tau} = \frac{\partial}{\partial x}\left(\alpha_m \frac{\partial u}{\partial x}\right) + \frac{\partial}{\partial x}\left(\alpha_m \delta \frac{\partial T}{\partial x}\right), \qquad (5.35)$$

where u is the percentage moisture content, α is the thermal diffusivity (m²/s), c_ρ is the specific heat (J/kg/K), λ is the specific heat of evaporation of water (J/kg), δ is the heat moisture conductivity coefficient (%/κ), α_m is the mass diffusivity for water (m²/s), Q is inlet heat source and ε is the phase-change criterion (dimensionless).

These equations form a nonlinear system because thermal and mass diffusion coefficients are functions of both moisture content and temperature.

The system expressed by Eqs. 5.34 and 5.35 has to be closed with boundary and initial conditions for temperature and moisture content. It is very complicate to find an analytical solution for this nonlinear boundary problem. Because of this, numerical solution techniques are very often used. In food and bioprocess modeling, finite-difference methods are commonly used especially in solving partial differential equations. There are some articles (e.g., Zheleva and Kambourova, 2001; Kambourova et al., 2003, 2004) where this system is solved numerically for baking or drying of dough food.

References

Anon., 1981. In the bag microwave processing. *Food Trade Review*, 51(3), 22–30.

Anon., 1984. Baking and ovens. History of heat technology. *Bakers Digest*, 58(2), 12–16.

Balaban, M., 1989. Effect of volume change in foods on the temperature and moisture content predictions of simultaneous heat and moisture transfer models. *Journal of Food Process Engineering*, 12, 67–88.

Charm, S.E., 1978. *The Fundament of Food Engineering*, 3rd ed., AVI, Westport, CT, Publishing Co.

Choi, Y., and Okos, M.R., 1986. Effects of temperature and composition on the thermal properties of foods. In *Food Engineering and Process Applications, Vol. 1, Transport Phenomenon*, L.Maguer and P.Jelen(eds.). Elsevier, New York, pp. 93–101.

Dickerson, R.W., 1969. Thermal properties of foods. In *The Freezing Preservation of Foods*, Vol.2, 4th ed., D.K.Tresseler, W.B.Van Arsdel, and M.J.Colpey(eds.). AVI, Westport, CT.

Earle, R.A. Unit Operation in Food Processing, http://www.nzifst.org.nz/unitoperations.

Encyclopedia Britannica, 2006. Food preservation, http://www.britannica.com (Accessed 3 Dec 2005, 15 May 2006).

Incropera, F.P. and De Witt, D.P., 1990. *Introduction to Heat Transfer*, 2nd ed., Wiley, New York.

Isachenko, V., Osipova, V. and Sukomel, A., 1981. Heat transfer, Moskow (in Russian).

Kambourova, V., Zheleva, I. and Mashkov, P., 2003. Optimization of impulse thermal regime for baking of bread with infrared heaters. *Comptes rendus de l'Academie bulgare des Sciences*, 56(12), 65–70.

Kambourova, V., Zheleva, I. and Mashkov, P., 2004. Mathematical modeling of temperature and moisture content of bread during baking in oven with infrared heating, *9th Workshop of Transport Phenomena in Two-Phase Flow, Borovets 2004, Proceedings*, pp. 197–204.

Knutson, K.M., Marth, E.H. and Wagner, M.K., 1987. Microwave heating of food. *Lebensmittel – Wissenschaft & Technologie*, 20, 101–110.

Lazar, M.E., Lund, D.B. and Dietrich, W.C., 1971. IQB – a new concept in blanching. *Food Trade Review*, 42(3), 13–15.

Luikov, A.V., 1975. Systems of differential equations of heat and mass transfer in capilary porous bodies. *International Journal of Heat and Mass Transfer*, 18, 1.

Martens, T. 1980. Mathematical model of heat processing in flat containers PhD thesis, Katholeike University of Leuven, Belgium.

Riedel, L., 1949. *Chemie Ingenieur Technik*, 21, 349.

Saltiel, C. and Datta, A.K., 1998. Heat and mass transfer in microwave processing. *Advances in Heat Transfer*, 33, 1–94.

Selman, J.D., 1987. The blanching process. In *Developments in Food Preservation-4*. Elsevier, London, pp.205–280.

Siebel, J.E., 1892. Specific heat of various products. *Ice and Refrigeration*, 2, 256–257.

Singh, R.P. and Heldman, D.R., 2001. Introduction to Food engineering, 3rd ed., Academic Press, London.

Sweat, V.E., 1986. Thermal properties of foods. In *Engineering Properties of Foods*, M.A.Raoand S.S.H.Rizvi (eds.). Dekker, New York

Sweat, V.E. and Haugh, C.G., 1974. A thermal conductivity probe for small food samples. *Transactions of the ASAE*, 17(1), 56–58.

Zheleva, I., Kambourova, V., 2001. Temperature and moisture content prediction in dough foods. *Theoretical and applied mechanics*, Proceedings of the 9th National Congress, Bulgaria, Vol. 2, pp. 82–88.

Chapter 6
Shelf-Life Prediction of Chilled Foods

Gudmundur Gudmundsson and Kristberg Kristbergsson

6.1 Introduction

All foods have a finite shelf life. Even foods, which mature with time, will in the end deteriorate, although their life span can exceed 100 years. Definitions of shelf life of food products differ. Some stress the suitability of the product for consumption, others for how long the product can be sold. The Institute of Food Science and Technology emphasizes safety in its definition of shelf life: "The period of time under defined conditions of storage, after manufacture or packing, for which a food product will remain safe and be fit for use" (http://www.ifst.org). This definition does not describe what makes a food product "safe" or "fit" for use, but one can say all factors which restrict the shelf life of a food product either affect safety or quality or both.

A definition of safety can be found on the Web site of the US National Library of Medicine and National Institutes of Health: "Food safety refers to the conditions and practices that preserve the quality of food to prevent contamination and food-borne illnesses" (http://www.nih.gov). There are many definitions on food quality depending on the field and perspective of the user of the term. Kramer and Twigg (1968) defined food quality "as the composite of characteristics which differentiate individual units and have significance in determining the degree of acceptability of that unit by the user."

Sufficient shelf life is important for food producers, retailers, and customers alike to decrease food deterioration and waste, but also to facilitate product development. Modern food production is characterized by the ever-increasing importance of product development, during which safety and quality have to be built into all new products. For these reasons considerable effort has been made by the food industry to increase the shelf life of food products, not least that of perishable products such as chilled foods. Through use of new technologies in processing and

K. Kristbergsson (✉)
Department of Food Science and Human Nutririon, University of Iceland, Hjardarhaga 2–6,
107 Reykjavik, Iceland

R. Costa, K. Kristbergsson (eds.), *Predictive Modeling and Risk Assessment,*
DOI: 10.1007/978-1-387-68776-6, © Springer Science + Business Media, LLC 2009

packing, the food industry has made significant progress in this respect. Nevertheless the shelf life of chilled foods can only be extended to a limited extent when compared with that of frozen or dried foods.

Since the determination of a product's shelf life is time-consuming and expensive, companies have been looking for methods to speed up the process and reduce costs. Shelf-life prediction, based on mathematical models, is one of the methods available to researchers to accelerate product development. The purpose of shelf-life prediction is to estimate the shelf life of prototypes of food products without carrying out direct storage tests. The benefits of shelf-life prediction are obvious. If a simple method or a model can be used to estimate the shelf life of a product, the costs of direct tests can be reduced and the phase of product development will be shortened.

The scope of this chapter is to describe methods and mathematical models used for shelf-life prediction of chilled foods.

6.2 Shelf Life of Chilled Foods

Chilled foods are foods which are stored refrigerated, but unfrozen. The purpose is to prolong shelf life and maintain safety and quality. Chilled foods are a very diverse class of foods. They may be classified according to the raw materials used, such as chilled dairy products, meat, seafood, and fruits and vegetables, or the classification may be based on the method of preparation, such as processed ready-to-eat foods, raw foods intended to be cooked prior to serving, and raw foods ready for consumption. New techniques in food processing, storage, and distribution have made new chilled food products possible and have extended the shelf life of others. The rapid increase in the availability of ready-made chilled foods is the result of new technology and so is the year-around supply of many exotic fruits and vegetables. In the latter case, the task of guarantying sufficient shelf life is exacerbated by long distribution lines.

In many countries regulations specify a maximum storage temperature for potentially hazardous chilled foods. In the UK, foods which need temperature control for safety should be stored at or above 63°C, or at or below 8°C (Office of Public Sector Information, 1995). In Australia, regulations require potentially hazardous foods to be kept at 5°C or below or at 60°C or above (Food Standards Australia New Zealand, 1998).

Information on the shelf life of chilled foods is usually advised by date marking. The EU general directive on food labeling requires most prepacked foods to carry such marking. This is normally "the date of minimum durability or, in the case of foodstuffs, which from the microbiological point of view are highly perishable, the 'use by' date" (European Commission, 2000). The "use by" date marking is used for perishable chilled foods, which require refrigeration to maintain their safety life rather than their quality. These are foods which can support the formation of toxins or growth of pathogens, and foods which are intended for consumption without sufficient treatment to destroy food-poisoning organisms. In the EU, some foods, such as fruits and vegetables, which have not been peeled or cut are exempt from date marking (European Commission, 2000).

6.2.1 Intrinsic and Extrinsic Factors

The shelf life of chilled foods is restricted by various mechanisms, which affect both the safety and the quality of the food products. Survival or growth of pathogens or spoilage bacteria will restrict the shelf life of many chilled foods, but chemical changes such as oxidation and physical changes such as moisture migration are also important.

The mechanisms affecting food are controlled by product-related intrinsic factors and process- or environment-related extrinsic factors (Table 6.1). The intrinsic factors include food composition and formulation, content and type of microorganisms, water activity, pH, and redox potential, and the extrinsic factors include food processing, storage temperature, relative humidity, packaging, and handling during retail operations and consumption.

6.2.2 Safety

For food products safety is more important than quality. In chilled foods, safety can be affected by changes in many different factors. Of these, survival and growth of pathogens are of most concern, but chemical changes such as oxidation or migration from food-contacting materials can also undermine the safety of chilled foods during their shelf life.

Table 6.1 Intrinsic and extrinsic factors affecting shelf life of chilled foods

Intrinsic factors	Extrinsic factors
Microbiological quality of raw materials	Good manufacturing and hygiene practices
Raw materials history	
Food formulation and composition	Hazard analysis critical control point
Food assembly and structure	Food processing
pH	Storage temperature
Type of acid present	Gas atmosphere
Water activity (a_w)	Relative humidity
Redox potential (E_h)	Packaging
Biological structures	Retail practices
Oxygen availability	Consumer practices
Nutritional content and availability	
Antimicrobial constituents	
Natural or artificial microflora of the food	

Reconstructed from Food Safety Authority of Ireland (2005), Jay (1996), and McDonald and Sun (1999)

Initial microbial, chemical, or physical contamination can and often do make a food product unfit for consumption, but a food product which is contaminated from the beginning has no shelf life. Shelf-life assessment is about *following changes* during storage and examining if critical parameters go from a safe to an unsafe level. Evaluation of these changes is carried out by microbiological, chemical, or physical methods depending on the factors jeopardizing safety. Sensory evaluation is obviously not used for monitoring changes in the safety of foods.

6.2.3   Quality

The factors affecting quality of chilled foods can be split into microbiological spoilage and chemical and physical changes. The combined effects of these changes will result in sensory alterations and in the end will make the product unacceptable to the consumer or the panel used to define the end point of organoleptic quality. This might be due to visual changes such as abnormal color or slime formation, or might be related to changes in smell, taste, or texture.

For many chilled foods microbiological spoilage is the most decisive factor restricting shelf life, but there are differences among different types of chilled foods, based on different intrinsic and extrinsic factors. Oxidation is important for products containing unsaturated lipids and in many products other chemical mechanisms such as hydrolysis or nonenzymatic processes speed up deterioration. The creaming of fat droplets in fresh milk is an example of a physical change which may restrict shelf life.

Although changes in sensory attributes are the result of chemical, microbiological, or physical changes, it is often difficult or impossible to measure the underlying mechanisms. As a result, sensory evaluation has to be used for monitoring food quality. Often a hedonic scale (0, 1, 2,...) is used to make data processing simpler.

A food's quality can be based on a minimum content of nutrients and certain levels of microorganisms, both of which go unnoticed by the consumer, but usually quality means sufficient sensory quality as judged by a panel or the consumers. The exact determination of sensory quality has to be carried out by a panel or by correlation to chemical, microbiological, or physical tests, which lead to loss of quality, but there are many practical problems in using sensory evaluation to monitor changes in chilled foods. One must differentiate between "end-point" information and quality-loss information.

6.3   Kinetics of Food Deterioration

"The key to the application of kinetics to prediction of quality loss is selection of the major mode of deterioration, measurement of some quality factor related to this mode, and application of mathematical models to make the needed predictions" (Labuza, 1984).

6.3.1 Deterioration of Quality

All chilled foods undergo deterioration during storage as a result of chemical and physical changes, or owing to the activity of microbes. These changes, separately or combined, lead to the deterioration of quality or decreased safety. The quality deterioration has been described by the following general equation:

$$dQ/dt = f(C_i, E_i) \tag{6.1}$$

where Q is quality, t is time, C_i is intrinsic factors of the product, E_i is extrinsic factors, and f is the function linking the quality change to C_i and E_i. The function can be of various types (Taoukis et al., 1997).

Foods are complex systems and many intrinsic and extrinsic factors are involved in quality deterioration, with different types of reaction mechanisms. If Eq. 6.1 is to be described in detail, it would be impracticable, with too many components, rate constants, and mechanisms. For simplicity, the changes taking place during quality deterioration can be presented by loss of a desirable quality factor (Q_{Ai}) or formation of an undesirable factor (Q_{Bi}) (Taoukis et al., 1997) and by ignoring the true mechanisms involved. The rates of change in Q_A and Q_B would be

$$-d[Q_A]/dt = k[Q_A]^n \tag{6.2}$$

and

$$d[Q_B]/dt = k'[Q_B]^{n'}, \tag{6.3}$$

where k and k' are *apparent* rate constants, and n and n' are the *apparent* orders of the reactions. Here only the concentration of factors Q_A or Q_B is considered, other factors being constant. In practice it is common to monitor the loss of only one major quality factor Q_A (or the gain of one undesirable factor Q_B). Since Eqs. 6.2 and 6.3 just differ in sign, the following discussion will be restricted to Eq. 6.2.

A word of caution is needed. The differential equation expressed in Eq. 6.2 is just a tool for making possible the extrapolation of the end-point value of Q_A and the estimation of the corresponding shelf life. It may have nothing to do with the real mechanisms and reactions. The equation is based on the assumption that all the reaction conditions remain constant.

6.3.2 Reaction Order

For most quality attributes in foods, the apparent order of the reactions described in Eq. 6.2 is zero, first, or second order (Taoukis et al., 1997) (Fig. 6.1). If the apparent order n is 0, Eq. 6.2 becomes

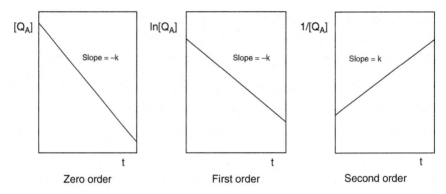

Fig. 6.1 Different reaction orders

$$-\mathrm{d}[Q_A]/\mathrm{d}t = k \qquad (6.4)$$

and after integration it becomes

$$[Q_A] = [Q_A]_0 - kt, \qquad (6.5)$$

where $[Q_A]_0$ is the initial concentration of Q_A. The rate of loss of $[Q_A]$ is linear and a plot of the loss of quality against time gives a straight line with slope $-k$ and intercept $[Q_A]_0$. Nonenzymatic browning is an example of a zero-order reaction (Taoukis et al., 1997; Gámbaro et al., 2006).

If n is equal to 1, Eq. 6.2 becomes

$$-\mathrm{d}[Q_A]/\mathrm{d}t = k[Q_A] \qquad (6.6)$$

and after integration

$$\ln[Q_A] = \ln[Q_A]_0 - kt. \qquad (6.7)$$

The rate depends on the quantity of Q_A which remains at time t. In this case a plot of $\ln[Q_A]$ against time gives a straight line with slope $-k$ and intercept $\ln[Q_A]_0$. This apparent reaction order is common in foods and fits many phenomena, including oxidative color loss, microbial growth and inactivation, vitamin loss, and texture loss in heat processing (Taoukis et al., 1997).

Quality deterioration with an apparent order of 2 is uncommon in foods, but an example is the initial loss of vitamin C in milk products (Labuza, 1984). In this instance Eq. 6.2 becomes

$$-\mathrm{d}[Q_A]/\mathrm{d}t = k[Q_A]^2 \qquad (6.8)$$

and after separation of variables and integration one obtains

$$1/[Q_A] = 1/[Q_A]_0 + kt. \tag{6.9}$$

Here a plot of $1/[Q_A]$ against time produces a straight line with slope k and intercept $1/[Q_A]_0$.

To be able to distinguish between zero- and first-order reactions it is necessary for the deterioration reactions to proceed far enough or at least to 50% conversion. This depends on the precision in measuring Q_A. If the precision is poor, an even larger change in Q_A is needed to distinguish between different orders, and to estimate the slope of the curve (Taoukis et al., 1997).

6.3.3   Arrhenius Equation

As stated before, the rate constant in Eq. 6.2 is an apparent and not a true constant. It changes with changes in intrinsic and extrinsic factors, such as temperature, moisture (or a_w), and pH. Of these factors, temperature is the most important.

The Arrhenius equation, which is derived from thermodynamics, describes the effect of temperature on the rate of chemical reactions:

$$k = A \exp(-E / RT), \tag{6.10}$$

where k is the rate constant, A is a constant, E is the activation energy (kJ/mol), R is the universal gas constant ($R = 8.314$ J/mol/K), and T is the absolute temperature (K). By definition the Arrhenius reaction is limited to an elementary process; however, it can often be used with success to describe an overall reaction. Then E becomes the *apparent* activation energy representing a combination of the activation energies of all the different reactions involved.

The activation energy, E, is a measure of the temperature sensitivity of the reaction and its value can be estimated by plotting experimental data for rate constants at different temperatures. E is 10–20 kcal/mol (approximately 40–80 kJ/mol) for simple hydrolysis reactions, 20–40 kcal/mol (80–170 kJ/mol) for nonenzymatic browning, and 50–150 kcal/mol (210–630 kJ/mol) for enzyme and microbial destruction (Labuza, 1984).

After integration, Eq. 6.10 can be rewritten as

$$\ln k = \ln A - E/RT. \tag{6.11}$$

A plot of $\ln k$ versus $1/T$ is called an Arrhenius plot and gives a straight line with a slope of $-E/R$ (Fig. 6.2).

The Arrhenius equation has been used to describe many other phenomena besides chemical reactions, including the effect of temperature on microbial growth and the kinetics of quality deterioration. For a change in quality factor Q_A where the apparent order is 0 and the relationship between the rate constant and temperature

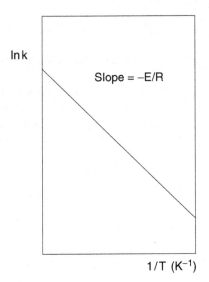

Fig. 6.2 Arrhenius plot

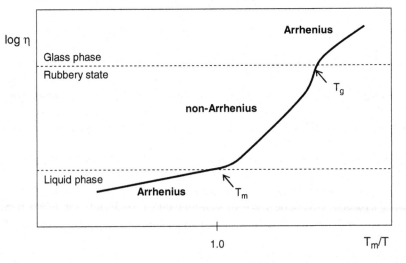

Fig. 6.3 Viscosity as a function of reduced temperature (T_m/T) for glassy and partially crystalline polymers. (Reconstructed from Levine and Slade 1987)

shows Arrhenius-type behavior, the differential equation relating changes in Q_A and time becomes

$$-d[Q_A]/dt = C \exp(-E / RT), \tag{6.12}$$

where C is a constant. If the apparent order of the reaction is 1, Eq. 6.10 becomes

$$-\mathrm{d}[Q_A]/[Q_A]=C\exp(-E/RT)\mathrm{d}t. \qquad (6.13)$$

Deviations from Arrhenius behavior are common and they can result from different factors. Changes in a_w or moisture can affect reaction rates through the glass transition. During rapid cooling, food polymers may enter the rubbery state between the melting point T_m and the glass-transition temperature T_g (Fig. 6.3). In the rubbery state, viscosity decreases and movements of molecules slow down. At the same time the slope of the Arrhenius plot may change and the material deviates notably from Arrhenius behavior (Taoukis et al., 1997). Changes in critical quality factors with temperature or a change of reaction mechanisms can also result in nonlinear Arrhenius plots.

6.4   Shelf-Life Testing

Shelf life of chilled foods can be estimated in various ways, by direct testing or by the use of external data. The data can be drawn from a data bank on pathogens and used for shelf-life prediction or they may be little more than "educated guesses" based on experience with similar products.

Direct storage tests include real-time storage trials and accelerated testing. Challenge tests are real-time tests and are normally used to assess microbial hazards or safety. Accelerated tests are used to assess quality deterioration.

6.4.1   Real-Time Storage Trials

Real-time testing is based on storage trials, carried out under realistic conditions. These trials give valuable results and a good estimation of a product's shelf life, but they take a long time and are expensive. Distribution abuse tests are one type of real-time test. They are important for checking if a product is usable or safe after going through the distribution cycle.

Throughout their life cycle, food products are not stored at a constant temperature, but at various temperatures. It is important to reflect this reality in shelf-life testing. For perishable foods, which are very sensitive to abuse handling, this is even more important. Temperature fluctuations after the product has left the store have also to be considered.

6.4.2   Microbial Challenge Tests

Microbial challenge tests are used to determine if one or more particular microorganisms can survive or grow in foods during processing and storage. The tests are normally carried out in laboratories under exactly defined conditions. Usually a unit of the food product is inoculated with the pathogen of concern or a suitable nonvirulent surrogate. After inoculation, the product is stored under controlled real-time conditions. If the product might be subject to temperature abuse, this should be included in the challenge study.

Challenge studies are very helpful in analyzing risks and minimizing hazards. They are very important for the validation of safety parameters for the shelf life of chilled foods, but challenge studies can not guarantee the safety of a product under commercial conditions.

Drawbacks of challenge tests are the costs involved and the time it takes to carry them out. Additionally, a challenge test is only valid for the exact conditions tested. If a product is reformulated, the test has to be repeated. Predictive models have turned out to be valuable in reducing the amount of challenge testing needed for chilled foods.

6.4.3 Accelerated Shelf-Life Testing

Accelerated shelf-life testing (ASLT) is based on storing a food product under some abuse condition, such as elevated temperature, and thereby speeds up deterioration and reduces the time necessary for testing. As a result, the process of shelf-life estimation can be shortened considerably compared with the time required for testing under normal conditions. ASLT is often used for foods which are stored at room temperature and contain unsaturated fat, but the method is also suitable for chilled foods.

ASLT has its drawbacks and may lead to erroneous estimation of shelf life. Storage at elevated temperature can change the mechanism of food deterioration or even change the critical factors which restrict the shelf life of a food product at normal storage temperature. At elevated temperature, reactions with high activation energies might predominate.

The possible changes in mechanisms and activation energy at high temperatures generate the need for upper temperature limits, above which it is unwise to do accelerated testing. The upper limits for ASLT of chilled foods have been estimated to be only 7–10°C (Labuza, 1984).

Instead of elevated temperature, changes in other extrinsic or intrinsic factors can be used to accelerate deterioration. These include changes in concentrations of reactants, water activity, and oxygen levels.

6.5 Mathematical Models

"The most common method [for shelf-life testing] is to select some single abuse condition such as high temperature, analyze the product two or three times during some specified storage period, and then extrapolate the results to normal storage conditions by using some "fudge factor", perhaps gained by experience with similar foods. The better approach is to assume that certain principles of chemical kinetics apply with respect to temperature acceleration, e.g., the Arrhenius relationship, and utilize kinetic design to produce a more accurate prediction" (Labuza, 1984).

During the last few decades mathematical models for shelf-life prediction of foods have gradually become available to food scientists and industry. These models range from simple equations to sophisticated software packages, but all serve two main purposes: one is to predict the shelf life of a product; the other is to aid in designing ways to inhibit microbial growth and increase the shelf life of products.

Mathematical models are often classified as mechanistic, empirical, or stochastic. Mechanistic models are based on fundamental reasoning and knowledge of the processes involved. Empirical models are based on a "black box" approach, where the objective is to get a convenient model for sets of data. The stochastic models are empirical models based on pre-established probability distributions. This classification appears to be somewhat arbitrary. The boundaries between the different types of models are not clear and all types have been used in shelf-life prediction of foods. Another classification appears to be more meaningful. The mathematical models are classified as primary or secondary models. *Primary models* describe growth, survival, or inactivation of microorganisms under constant conditions. *Secondary models* describe how parameters from primary models behave under a range of conditions.

Mathematical models for prediction of the shelf life of foods can be very valuable, but they should always be used with caution. The models are to be used as guidance during product development to reduce the amount of testing by reducing the number of prototypes put forward for shelf-life testing. The final formulation should always be validated adequately and safety demonstrated by challenge studies.

6.5.1   Predictive Microbiology

Predictive microbiology is concerned with the mathematical description of microbial growth and survival in foods. It is a powerful tool for evaluating microbial hazards and premature spoilage of foods during storage. By the use of predictive models, the shelf life of foods differing slightly in formulation, processing, or packing can readily be estimated. (see Chap. 3, where predictive microbiology is covered in detail).

Temperature fluctuations make modeling of microbial growth and inactivation demanding. From this stems the "history effect," according to which the rate of quality loss at time t is a function of prior storage time and temperature. Temperature changes are typically random and therefore it is difficult to give a mathematical description of the time–temperature variation (Koutsoumanis, 2001).

For chilled foods, microbial spoilage and microbial hazards are normally most important and shelf-life predictions of these deal mainly with microbiological aspects. For these foods the most important microbes are those which may survive or grow at refrigeration temperatures. These may be both spoilage bacteria and pathogens but since safety is more important than quality, possible survival or growth of pathogens is of primary importance. *Listeria monocytogenes* may grow at refrigeration temperature and this pathogen is a threat in many chilled products. In cooked ready-made foods the spore-forming bacteria *Clostridium botulinum* and *Bacillus cereus* are of great concern. These chilled foods are usually minimally processed, low in additives, and rely on refrigeration for preservation.

6.5.2 Nonmicrobial Changes

Important types of chemical deterioration are lipid oxidation, enzymatic degradation, and nonenzymatic browning. Loss of nutrients and changes in color are also a result of chemical changes.

Physical deterioration factors in refrigerated foods include phase changes such as freezing of water, migration of moisture, glass transitions, and phase separation. Glass transition affects the quality of foods in many ways. If the temperature or the moisture content is high enough, foods may pass from glass and enter the rubbery state and as a result lose crispiness owing to texture changes. The coalescence of fat droplets in milk and other food emulsions is a physical change which can limit the shelf life of products. These changes can be described by the Stokes equation:

$$v_s = \alpha(\rho_D - \rho_D)d^2/18\eta_C, \tag{6.14}$$

where v_s is the rate of creaming, α is acceleration, ρ_D is the density of the dispersed phase, ρ_C is the density of the continuous phase, d is the diameter of the droplets, and η_C is the viscosity of the continuous phase (Walstra, 1996).

Temperature fluctuations make modeling of quality deterioration a challenge for food scientists. Corradini and Peleg (2006) pointed out "the logarithmic degradation rate of a complex biochemical process isothermal as well as nonisothermal, may depend not only on temperature (and other factors, like pH, etc.) but also on the system's momentary state."

6.5.3 The Critical Factor

The evaluation of quality deterioration in foods is usually complex and it is hard to define an exact end point. In chilled foods microbiological considerations are most important for safety considerations, but when dealing with quality, chemical, physical, sensory, and microbiological considerations may all play a part. Despite the complexity of food deterioration, it is often possible to pinpoint one or very few quality factors which can be used to determine end points of shelf life. The critical factors are obviously different between different food products, depending on intrinsic and extrinsic factors.

Mathematical predictions of shelf life are usually based on one critical quality factor, which is modeled. In foods with two or more critical quality factors, with different activation energies, the shelf-life prediction can be very complex.

6.5.4   Examples of Mathematical Models

6.5.4.1   Arrhenius Models

Models based on the Arrhenius equation have been used for describing both microbiological growth and quality deterioration. The Arrhenius plot often exhibits deviations at low temperatures, which may limit its use at refrigeration temperatures. Davey (1993) showed that these deviations can be accounted for in an additive Arrhenius model:

$$\ln k = C_0 + C_1/T + C_2/T^2, \tag{6.15}$$

where k is the microbial growth rate and C_0, C_1, and C_2 are constants. The term C_2/T^2 accounts for the curve in the plot. Davey (1993) also developed an Arrhenius model where a_w was included.

6.5.4.2   Baranyi and Roberts Model

The model of Baranyi and Roberts (1994) is a relatively complicated primary model, and has enjoyed popularity in predictive microbiology:

$$y(t) = y_0 + \mu_{max} A(t) - 1/m[\ln(1 + \{\exp[m\mu_{max} A(t)]$$
$$- 1\}/\exp[m(y_{max} - y_0)])] \tag{6.16}$$

where $y(t) = \ln x(t)$, where x is the cell concentration at time t; $y_0 = \ln x(t_0)$, where x is the cell concentration at time t_0; $y_{max} = \ln x_{max}$, where x_{max} is the maximum cell concentration; μ_{max} is the maximum specific growth rate; and m is a curvature parameter. $A(t)$ is a function which plays the role of a gradual delay in time:

$$A(t) = t + \{\ln[\exp(-\mu_{max}t) + \exp(-h_0) - \exp(-\upsilon t - h_0)]\}/\mu_{max}, \tag{6.17}$$

where $h_0 = -\ln \alpha_0$, υ is rate of transition from the lag to the exponential phase, and

$$\alpha_0 = z_1(t_0)/[K_z + z_1(t_0)]. \tag{6.18}$$

The parameter α_0 is called the physiological state of the cells at time t_0 and $z_1(t_0)$ is the quantity of a critical substance causing bottleneck in the growth at time t_0. K_z is a parameter depending on those external variables which are independent of the growing culture (Baranyi and Roberts, 1994, 1995).

6.5.4.3 Square-Root Models

Square-root models are secondary models which can be used to describe the rate of microbial growth as a function of temperature (Ross, 1993):

$$\sqrt{k} = b(T - T_{min})\{1 - \exp[c(T - T_{max})]\}, \qquad (6.19)$$

where k is the microbial growth rate, b and c are constants, and T_{min} and T_{max} are the model's minimum and maximum temperatures of growth for the relevant microorganism and not the actual minimum and maximum growth temperatures. For temperatures below the optimum temperature, a simpler approximation can be used:

$$\sqrt{k} = b(T - T_{min}). \qquad (6.20)$$

The square-root models can be easily modified to include in addition to temperature other factors such as a_w and pH (Ross, 1993). According to Ratkowsky (1993), the estimates of the parameters are "close to being unbiased, normally distributed, minimum variance estimators."

Square-root models have also been used to describe quality deterioration at refrigeration temperature, if the principal microbe of spoilage and its T_{min} is known (McMeekin and Ross, 1996). The models have been put to use in the concept of relative rate, which is used in the relative rate of spoilage (RRS) models such as in the Seafood Spoilage Predictor (Dalgaard et al., 2002).

6.5.4.4 Relative Rate of Spoilage Models

RRS models are used for the calculation of shelf life at one temperature compared with a preferred or reference temperature. RRS models are often based on the square-root model:

$$\text{Relative rate} = (T - T_{min})^2 / (T_{ref} - T_{min})^2, \qquad (6.21)$$

where T is the temperature of interest and T_{ref} is the preferred storage temperature. T_{min} is the model's minimum temperature of growth for the microbe responsible for spoilage (Ross, 1993). RRS models allow shelf life to be predicted at different temperatures although the shelf life has been determined by sensory evaluation at only a single storage temperature (Dalgaard et al., 2002).

6.5.4.5   Logistic Models

The logistic models are primary models, which are based on probability theory. They have been used for describing microbial growth:

$$\log(N_t) = \log(N_{max}/\{1 + [(N_{max}/N_0)1]\exp(-\mu_{max}t)\}), \tag{6.22}$$

where N_t is the number of microorganisms at time t, N_{max} is the maximum number of microorganisms, N_0 is the initial number at time zero, and μ_{max} is the maximum specific growth rate. In the form above, the logistic model describes microbial growth as a sigmoid curve without a lag phase (Dalgaard et al., 1997).

6.5.4.6   Weibull Models

Weibull models are special cases of the more general power-law models. They are primary models and are based on the Weibull probability density function (van Boekel, 2002):

$$f(t) = \beta/\alpha(1/\alpha)^{\beta-1}\exp[-(t/\beta)^{\beta}], \tag{6.23}$$

where t, α, and β are constants and are greater than zero. The Weibull distribution function is used in describing failure phenomena in engineering as well as inactivation in predictive microbiology (van Boekel, 2002). The Weibull models are very versatile.

Jagannath et al. (2005) used a two-parameter Weibull equation to describe inactivation of spores of *Bacillus subtilis:*

$$\log N = \log N_0(t/\delta)^p, \tag{6.24}$$

where N is the number of spores at time t, N_0 is the number of spores at time zero, δ is the reciprocal of the growth rate, and p is a shape parameter.

Corradini and Peleg (2006) used a Weibull power-law model to describe the kinetics of isothermal degradation of vitamins:

$$C_t/C_0 = \exp[-bt^n], \tag{6.25}$$

where C_t is the concentration at time t, C_0 is concentration at time zero, b is a temperature-dependent coefficient, and n is a fixed average power (Corradini and Peleg, 2006).

Survival analysis based on Weibull models has been used in shelf-life prediction (Hough et al., 2003; Gámbaro et al., 2006).

6.5.4.7   Williams, Landel, and Ferry Models

Williams, Landel, and Ferry (WLF) models are based on the WLF equation. This is an empirical equation which describes the temperature dependence of various phenomena in amorphous polymers at the rubbery state above the glass-transition temperature. The equation has been used to describe the deterioration rate in amorphous foods in the rubbery state (Slade and Levine, 1994; Ashokan and Kokini, 2005).

The general form of the WLF equation is

$$\log \alpha_{\mathrm{T}} = -C_1(T - T_g)/(C_2 + T - T_g), \qquad (6.26)$$

where α_{T} is the ratio of the relaxation phenomenon at temperature T to the relaxation at the glass-transition temperature T_g and C_1 and C_2 are coefficients which depend on the system under study (Slade and Levine, 1994; Ashokan and Kokini, 2005). Instead of using T_g, one can use a reference temperature T_{ref} above T_g. The coefficients can then be adjusted to T_g according to

$$C_1 = C_{1ref} C_{2ref}/(C_{2ref} + T_g - T_{ref}) \qquad (6.27)$$

and

$$C_2 = C_{2ref} + T_g - T_{ref}, \qquad (6.28)$$

where T_{ref} is the reference temperature and C_{1ref} and C_{2ref} are coefficients at T_{ref} (Ashokan and Kokini, 2005).

6.5.5   Statistical Considerations

Adequate experimental design and good data are the basis of every good model of shelf-life prediction, but in addition to normal good laboratory practices in data handling, the predictive models chosen have to be validated and their goodness of fit inspected. Predictive models are usually based on data from liquid media and there is no guarantee that predicted values from a model will fit the special food of interest.

Ratkowsky (1993) pointed out that almost all predictive models are nonlinear, and therefore "it is important to be able to ascertain the extent to which the parameter estimators are biased and non-normally distributed." Ratkowsky (1993) listed five points which scientists should be aware of when dealing with nonlinear regression models:

1. Parsimony (models should contain as few parameters as possible)
2. Parameterization (find the one which has the best estimation properties)
3. Range of applicability (the data should cover the full range of X and Y)
4. Stochastic specification (the error term needs to be modeled too)
5. Interpretability (parameters should be meaningful, as far as possible)

Testing of a model's goodness of fit consists of inspection of residuals and calculation of the coefficient of correlation, R^2. The residuals are inspected by comparing predicted and observed values and checking if they are evenly distributed around line of symmetry. Caution is needed when using the coefficient of correlation R^2, because, generally, the fit of a model improves when terms or parameters are added (Davey, 1993; Baranyi and Roberts, 1995). Adding additional terms to a model clearly goes against parsimony, which is one of the criteria for good models. Instead of calculating R^2, Davey (1993) suggested the use of percentage variance accounted for ($\%V$). The association between the two is shown below:

$$\%V = [1 - (1 - R^2)(n-1)/(n-1-N_T)] \times 100, \qquad (6.29)$$

where n is the number of data and N_T is the number of terms. It can be seen from Eq. 6.29 that when n increases, $\%V$ approaches $R^2 \times 100$ (Davey and Amos, 2002).

6.5.6   Databases and Software

There are many types of databases and software available to food scientists working on shelf-life prediction of chilled foods. All the models have in common that they have to be used conservatively and in combination with validation tests.

ComBase, or the "Combined Database of Microbial Responses to Food Environments," is a large international database for predictive food microbiology accessible on the Internet. ComBase is a consortium of the Food Standards Agency and the Institute of Food Research, UK, the Agricultural Research Service of the United States Department of Agriculture (USDA-ARS) and its Eastern Regional Research Center, and the Australian Food Safety Centre of Excellence. Access to ComBase is free of charge and the homepage of the database is at http://www.combase.cc/.

The Pathogen Modeling Program is from the USDA-ARS. It is a package of models for predicting growth, survival, and inactivation of pathogens and toxin-producing bacteria in foods. The program can readily be used in shelf-life prediction of refrigerated foods. The Pathogen Modeling Program is available free of charge and can be downloaded from the Internet from the Web page of the USDA-ARS at http://ars.usda.gov/Services/.

Seafood Spoilage Predictor is a software program for predicting the shelf life of seafood at various temperatures. By combining data from temperature loggers, the software can predict the effect of temperature on shelf life. Seafood Spoilage Predictor contains two types of RRS models: one is a square-root model and the other is an exponential model (Dalgaard et al., 2002). Seafood Spoilage Predictor is freeware and can be downloaded from http://www.dfu.min.dk/micro/ssp/.

Growth Predictor and Perfringens Predictor are software programs for predicting growth of microorganisms under different conditions. Growth Predictor contains growth models for various microorganisms, but Perfringens Predictor provides models for predicting the growth of *Clostridium perfringens* during cooling of meats. The Food Standards Agency and the Institute of Food Research in the UK are responsible for the software, which is available free of charge from http://www. ifr.ac.uk/Safety/GrowthPredictor/.

Microfit is a software tool which was developed by the Institute of Food Research in the UK. The tool can be used for extracting microbial growth parameters from challenge tests and to fit growth models to data. Microfit is available free of charge from http://www.ifr.ac.uk/MicroFit/.

Food Spoilage Predictor© uses integration of data loggers and mathematical models for predicting remaining shelf life in chilled food products. Food Spoilage Predictor© was developed at the Department of Agricultural Science, University of Tasmania, Australia. The software is available from Hastings Data Loggers, Australia. Information on the software can be found at http://www.arserrc.gov/cemmi/FSPsoftware.pdf.

References

Ashokan, B.K. and Kokini, J.L. (2005): "Determination of the WLF constants of cooked soy flour and their dependence on the extent of cooking," *Rheologica Acta*, 45(2), 192–201.

Baranyi, J. and Roberts, T.A. (1994): "A dynamic approach to predicting bacterial-growth in food," *International Journal of Food Microbiology*, 23(3–4), 277–294.

Baranyi, J. and Roberts, T.A. (1995): "Mathematics of predictive food microbiology," *International Journal of Food Microbiology*, 26(2), 199–218.

Corradini, M.G. and Peleg, M. (2006): "Prediction of vitamins loss during non-isothermal heat processes and storage with non-linear kinetic models," *Trends in Food Science & Technology*, 17(1), 24–34.

Dalgaard, P., Mejlholm, O., and Huss, H.H. (1997): "Application of an iterative approach for development of a microbial model predicting the shelf-life of packed fish," *International Journal of Food Microbiology*, 38(2–3), 169–179.

Dalgaard, P., Buch, P., and Silberg, S. (2002): "Seafood Spoilage Predictor – development and distribution of a product specific application software," *International Journal of Food Microbiology*, 73(2–3), 343–349.

Davey, K.R. (1993): "Linear-Arrhenius models for bacterial growth and death and vitamin denaturations," *Journal of Industrial Microbiology*, 12(3–5), 172–179.

Davey, K.R. and Amos, S.A. (2002): Untitled (Letters to the Editor), *Journal of Applied Microbiology*, 92(3), 583–587.

European Commission (2000): "Directive No 2000/13/EC of 20 March 2000 on the approximation of laws of the Member States relating to the labelling, presentation and advertising of foodstuffs," Official Journal L Series 109, 06/05/2000, Brussels, 29. Downloaded from http://europa.eu.int/eur-lex/ on 1 August 2006.

Food Safety Authority of Ireland (2005): "Guidance Note No. 18 Determination of Product Shelf-Life," Food Safety Authority of Ireland, Dublin, p. 13.

Food Standards Australia New Zealand (1998): "Standard 3.2.2 Food Safety Practices and General Requirements," Division 1, Clause 1. Downloaded from http://www.foodstandards.gov.au/ on 1 August 2006.

Gámbaro, A., Gastón, A., and Giménez, A. (2006): "Shelf-life estimation of apple-baby food," *Journal of Sensory Studies*, 21(1), 101–111.

Hough, G., Longhor, K., Gómez, G., and Curia, A. (2003): "Survival analysis applied to sensory shelf life of foods," *Journal of Food Science*, 68(1), 359–362.

Jagannath, A., Tsuchido, T., and Membré, J.-M. (2005): "Comparison of the thermal inactivation of *Bacillus subtilis* spores in foods using the modified Weibull and Bigelow equations," *Food Microbiology*, 22(2–3), 233–239.

Jay, J.M. (1996): "Intrinsic and extrinsic parameters of foods that affect microbial growth." In: Jay, J.M.(ed.), Modern Food Microbiology, 5th edition, Chapman & Hall, New York, pp. 38–66.

Koutsoumanis, K. (2001): "Predictive modeling of the shelf life of fish under nonisothermal conditions," *Applied and Environmental Microbiology*, 67(4), 1821–1829.

Kramer, A. and Twigg, B. (1968): "Measure of frozen food quality and quality changes," In: Tressler, D.K. (ed.), The Freezing Preservation of Foods, 4th edition, Vol. 2, AVI Publishing Company, Inc., Westport, CT, pp. 52–82.

Labuza, T.P. (1984): "Application of chemical kinetics to deterioration of foods," *Journal of Chemical Education*, 61(4), 348–358.

Levine, H. and Slade, L. (1987): "Water as a plasticizer: physico-chemical aspects of low-moisture polymeric systems." In: Franks, F. (ed.), Water Science Reviews, Vol. 3, Cambridge University Press, Cambridge, pp. 79–185.

McDonald, K. and Sun, D.-W. (1999): "Predictive food microbiology for the meat industry: a review," *International Journal of Food Microbiology*, 52(1–2), 1–27.

McMeekin, T.A. and Ross, T. (1996): "Shelf life prediction: status and future possibilities," *International Journal of Food Microbiology*, 33(1), 65–83.

Office of Public Sector Information (1995): Statutory Instrument 1995 No. 2200, "The Food Safety (Temperature Control) Regulations 1995," part II. Downloaded from http://www.opsi.gov.uk/ on 1 August 2006.

Ratkowsky, D.A. (1993): "Principles of nonlinear regression modeling," *Journal of Industrial Microbiology*, 12(3–5), 195–199.

Ross, T. (1993): "Bĕlehrádek-type models," *Journal of Industrial Microbiology*, 12(3–5), 180–189.

Slade, L. and Levine, H. (1994): "Water and the glass transition – dependence of the glass transition on composition and chemical structure: special implications for flour functionality in cookie baking," *Journal of Food Engineering*, 22 (1–4), 143–188.

Taoukis, P.S., Labuza, T.P., and Saguy, I.S. (1997): "Kinetics of food deterioration and shelf-life prediction." In: Valentas, K., Rotstein, J., Singh, E.R.P. (eds), The Handbook of Food Engineering Practice, CRC Press, New York, pp. 363–405.

van Boekel, M.A.J.S. (2002): "On the use of the Weibull model to describe thermal inactivation of microbial vegetative cells," *Journal of International Food Microbiology*, 74(1–2), 139–159.

Walstra, P. (1996): "Dispersed systems: basic considerations." In: Fennema, O.R. (ed.), Food Chemistry, 3rd edition, Marcel Dekker, Inc., New York, pp. 95–155.

Chapter 7
Exposure Assessment of Chemicals from Packaging Materials

Maria de Fátima Poças and Timothy Hogg

7.1 Introduction

A variety of chemicals may enter our food supply, by means of intentional or unintentional addition, at different stages of the food chain. These chemicals include food additives, pesticide residues, environmental contaminants, mycotoxins, flavoring substances, and micronutrients. Packaging systems and other food-contact materials are also a source of chemicals contaminating food products and beverages. Monitoring exposure to these chemicals has become an integral part of ensuring the safety of the food supply. Within the context of the risk analysis approach and more specifically as an integral part of risk assessment procedures, the exercise known as exposure assessment is crucial in providing data to allow sound judgments concerning risks to human health. The exercise of obtaining this data is part of the process of revealing sources of contamination and assessing the effectiveness of strategies for minimizing the risk from chemical contamination in the food supply (Lambe, 2002).

Human exposure to chemicals from food packaging and other food-contacting materials may occur as a result of migration from the packaging materials into the foods and beverages. The extent of migration and the inherent toxicity of the substance in question are the two factors which define the human health risk represented by packaging materials. In a formal risk analysis context the key components to be considered in a risk assessment of a packaging material are (1) chemistry and concentration data of the substance (exposure assessment) and (2) toxicology data (hazard characterization). In exposure assessment the use and the intended technical effect of the substance in packaging must be identified, the analytical methods for detection and/or quantification of the substance in the foods and in the packaging itself must be identified and implemented, and data for migration from packaging into foods and an evaluation of the consumer food intake must be collected. The hazard characterization component includes toxicology studies and the effects of different levels on

M. de Fátima Poças (✉)
Packaging Department, College of Biotechnology, Catholic University of Portugal, Rua Dr. António Bernardino de Almeida, 4200-072, Porto, Portugal

R. Costa, K. Kristbergsson (eds.), *Predictive Modeling and Risk Assessment,*
DOI: 10.1007/978-1-387-68776-6, © Springer Science+Business Media, LLC 2009

health, and a comprehensive profile of the substance, including possible decomposition products.

Exposure assessment is defined by WHO (1997) as the qualitative and/or quantitative evaluation of the likely intake of biological, chemical, or physical agents via food as well as exposure from other sources if relevant. As briefly outlined, it is one of the key parts of the risk assessment process. The focus of this chapter is on methods and mathematical approaches to assess the consumer exposure to chemicals migrating from food packaging systems.

7.2 Methods for Exposure Assessment

The choice of method for carrying out an exposure assessment is influenced by the purpose of the exercise, by the nature of the chemical, and by the resources and data available (Lambe, 2002). When estimating chemical exposure, four basic guiding principles must be followed (Rees and Tennant, 1993):

- The estimate should be appropriate for the purpose.
- The estimate should have an assessment of accuracy.
- Any underlying assumptions should be stated clearly.
- Critical groups of the population should be taken into account when these groups are disproportionally affected by the chemical.

The differences between food additives (direct) and substances migrating from packaging (referred to in US context as indirect additives) are such that different methods are necessary to prepare estimates of consumer exposure. When the substance in question is not a direct food chemical, such as a food additive, a natural toxin, or a pesticide residue, for example, but a chemical migrating from the packaging system, additional information is needed on the nature and composition of the packaging materials, the types of packages used for certain foods (related to packaging usage factors), and data on migration levels of the substances.

Exposure, in a dietary context, is defined as the amount of a certain substance that is consumed (Holmes et al., 2005) and is usually expressed as amount of substance per mass of consumer body weight per day. The general model to describe the exposure to chemicals from food packaging can be represented as

$$\text{Exposure} = \text{migration}\,(C_0, A/V, t, T, \ldots) \times \text{food consumption}, \qquad (7.1)$$

where "migration" represents the amount of chemical migrating into the food. The migration level depends on several variables, such as the packaging material itself, the chemical and physical nature of the food in contact, the initial concentration of the substance in the packaging material (C_0), time (t), and temperature (T), and it also depends on the ratio of the surface area of the packaging material (A) to the amount of food product (V). Food consumption represents the intake of food packed in a certain type of packaging system that has the migrating chemical.

The methods followed for assessing direct food chemicals, such as food additives, contaminants, and natural toxins and residues of pesticides or veterinary drugs, have been widely reported (Kroes et al., 2002; Luetzow, 2003). They are usually distinguished between:

- Point estimates – that use a single "best guess" estimate for each variable within a model (Vose, 2000).
- Probabilistic analysis – that involves describing variables in terms of distributions to characterize their variability and/or uncertainty. In some cases, it uses distributions of food intake but a fixed value for the concentration.

Point estimates are often used in a screening phase of exposure assessments with conservative estimates of variable and uncertain inputs to ensure "worst-case" or upper-bound estimates of exposure (Hart et al., 2003). These analyses are usually quick, cheap, and can serve to identify potential exposures that are so low that more detailed analyses are not worthwhile. However, owing to variability and uncertainty in the model input variables, deterministic point estimates provide no indication of the magnitude of uncertainty surrounding the quantities estimated and provide no insight into the key sources of underlying uncertainty, although some insight can be provided by deterministic sensitivity analyses (Kroes et al., 2002).

Probabilistic analysis gives quantitative information about both the range and the likelihood of possible exposure values for a set of individuals, allowing for the characterization of variability and uncertainty in model outputs. Additionally, it is possible to identify key sources of uncertainty and variability that can be the focus of future data collection, research, or model development efforts (Kroes et al., 2002; Cullen and Frey, 1999).

The methods can be characterized by a hierarchical and stepwise approach, proceeding from the level of more conservative assumptions and lower accuracy to the more refined estimates if the less exact results do not rule out the possibility of compliance with the food safety objective. The EU proposed a tiered approach (EC, 2001) for monitoring of food additive intakes (Table 7.1).

In assessing exposure to chemicals from packaging materials, a similar tiered approach for combining or integrating the food consumption data with the migration data may be applied. These approaches may range from making the assumptions of 100% migration, to more refined methods taking into account quantitative measures of migration into foods linked to survey data on intake of specific foods.

Table 7.1 European proposed tiered approach for exposure assessment of food additives

	Additive level or concentration	Food consumption data
Tier 1	Maximum permitted usage levels	Theoretical food consumption
Tier 2	Maximum permitted usage levels	Actual national food consumption
Tier 3	Actual usage levels	Actual national food consumption

7.3 Data Collection for the Assessments

As shown in Eq. 7.1, two types of data are required to estimate the exposure or the intake of a substance – migration and food consumption. This section briefly presents the techniques for data collection.

7.3.1 Migration Data

Migration data may be obtained from monitoring levels of chemicals in real food systems. There are major analytical difficulties with this practice owing to the complexity of the food matrixes and owing to chemical instability of some migrants. More commonly, migration data are obtained from migration experiments, carried out under certain sets of conditions of time and temperature of contact between the materials and the food, and using the food simulants instead of the food itself: water, ethanol solutions, acetic acid solutions, and olive oil (Directives EEC 85/572, EC 97/48). Recently, a diffusion model for estimation of the migration of components from plastic materials has been allowed for determining compliance with regulations (Directive 2002/72/EC).

Food consumption and packaging usage data are, unfortunately, rarely collected together, as most of the surveys are designed to gather information on energy, nutrients, or certain residues and food contaminants or additives regardless of the packaging system the food was contained in. Thus, adequate information on packaging usage is not easily available. For assessing the exposure to packaging chemicals, it is necessary to know what type of food (chemical and physical nature) is packaged in what type of material, as this determines the presence and concentration of the chemical and influences the potential for migration into the type of food. For example; considering the consumption of mineral water, the chemicals migrating from poly(ethylene terephthalate) bottles are certainly different from those migrating into the same mineral water packaged in polycarbonate bottles. From a different point of view, if the migration from poly(ethylene terephthalate) bottles is considered, the migration value of a certain substance will be different if we consider a soft drink (acid in nature) instead of a mineral water (Poças and Hogg, 2007).

Another factor to be considered in linking packaging usage, food consumption, and migration data is the ratio of the surface area of the packaging to the volume of the food product. Mass transfer of the chemical is a surface-area phenomenon and the concentration achieved in the food or food simulant depends on its volume or mass. Therefore, the size and the format of the packages are also very important (Poças and Hogg, 2007).

The approach for collecting statistics on food-packaging materials usage is very much an iterative process whereby data obtained from different complementary sources are processed to generate an overview. Data collection may be based on industry sources and on shelf audits along with discussions with food producers, packaging converters, raw material suppliers, and finally with trade associations, etc. The packaging statistics must be collected segmented into food type, pack size, and packaging material. For each of these, the number of units of consumption, surface area, weight of packaging, and coating thicknesses must be estimated (ILSI, 2001).

7.3.2 Food Consumption Data

Much more information is available for food consumption data. The method for collection is directly related to the purpose of the exposure assessment and to the accuracy of the model. Several methods can be used to estimate the intake of a food chemical and the choice will depend on what information is available and how accurate and detailed the estimate needs to be (Kroes et al., 2002).

Food consumption data may be collected at different levels: national/regional, household, or individual. When exposure in specific subgroups (e.g., children) is of interest, data on an individual level are essential. Table 7.2 presents some characteristics of the common methods for data collection at different levels.

Since 1993, an effort has been made to create a cost-effective, European databank based on food, socioeconomic, and demographic data from nationally representative household budget surveys through the DAFNE project– Data Food Networking. The DAFNE databank includes data from 44 household budget surveys of eight EU member states and Norway. The year of and the procedures for data

Table 7.2 Methods for the collection of food consumption data

Level	Method of data collection
National	Food balance sheets: • Results are an estimate of the average per capita value
Household	Data are collected by record keeping, interviews, or both, for a period of 2 weeks usually: • Budget surveys that gives information on the purchases in terms of expenditure or consumption surveys recording the amounts of food and drink brought into the household • Provides estimates of mean total intake of foods • Lacks data on the individuals consumption of foods • Possible overestimation of the intake owing to waste and underestimation of intake for foods consumed outside the home
Individuals	Data are collected by: • Twenty-four-hour recalls where the subject is asked to recall and describe the kinds and amounts of food and drink ingested during the immediate past period • Food frequency questionnaire consisting of a list of individual foods or food groups to assess the frequency of consumption • Food or dietary records kept for a specific time period • Diet history, where a trained interviewer assesses an individual's total usual food intake and meal pattern • Provides information on average food intake and distribution over various groups of individuals

collection and storage, the methodological characteristics, and other general information regarding the integrated datasets are presented in the project report. The tasks undertaken in the context of the DAFNE III project included the (1) incorporation of the raw household budget surveys data of each participating country in the central database, operating at the coordinating center, (2) harmonization of the food, demographic, and socioeconomic information collected in the surveys of the nine participating countries, and (3) estimation of the average daily food availability for the overall population and sociodemographic groups.

This initiative, allowing for monitoring dietary patterns, is very important for the improvement of health and related information available in the EU member states and for the development and improvement of systems of data collection, since methods can be harmonized. However, the majority of consumption surveys are designed for nutritional purposes and it is the food consumed which is important rather than the packaging of the food. When using these surveys for other purposes, such as the assessment of exposure to migrating chemicals from packaging materials, one should take the methodological inadequacies of this approach into account.

7.4   Point Estimates and Screening Approaches

There are a number of factors that influence the choice of model for any given exposure assessment. A stepwise approach with regard to the accuracy of the results is advised by WHO: from initial screening methods, to more specific methods that use actual concentration and consumption data, and in some cases more accurate methods to confirm previous results, such as the use of biomarkers or duplicate diet studies. Screening methods are usually based on assumptions for both model variables (food consumption or packaging usage and migration), leading to a considerable overestimation of exposure. This estimate may allow for the determination as to whether any further work, using more sophisticated models to provide results with higher certainty, might be needed. If this initial screening indicates that there is no practical likelihood that consumers are exposed to levels exceeding the safety objectives, for example, the tolerable daily intake (TDI) or the acceptable daily intake, or if the intake is clearly below an agreed threshold of toxicological concern, it may be decided that no further refinement of the assessment is needed (Kroes et al., 2002).

The screening phase of exposure assessment is very often, for direct food chemicals, performed as point estimates since they are relatively simple and inexpensive to carry out. Point estimates consider a single value for consumption of a food (usually the mean population value) and a fixed value for the chemical concentration in that food. For packaging-migrating chemicals this single value can represent:

- The value assuming that all of the substance initially present in the packaging material migrates into the food product (see Examples 1 and 2)
- The value corresponding to the specific migration limit (legal limit)
- An average migration value obtained experimentally (see Example 3)

Inherent to the point estimate approach are the assumptions that all individuals consume the specified food(s) at the same level, that the food chemical is always present in the food(s), and that it is always present at an average or high level. This approach does not provide insight into the range of possible exposure that may occur within a population, or the main factors influencing the results of the assessment (Kroes et al., 2002).

Example 1 Concentration of the substance in the food

A film of low-density polyethylene with a thickness of 100 μm is used for sandwich bags with fatty substances at the food surface. The material has an antioxidant additive incorporated into it – Irganox 1076 [octadecyl 3-(3,5-di-*tert*-butyl-4-hydrxyphenyl) propionate] at a concentration of 500 mg/kg. The bags have sides of 15 cm and the amount of product corresponds to a volume of 750 cm³.

The value assuming 100% of migration is calculated as

$$\text{Total migration} = \frac{\text{initial concentration} \times \text{package surface area} \times \text{thickness}}{\text{product}}$$

$$= \frac{2 \times 15 \times 15 \times 100E(-4) \times 500}{750} = 3\,mg/kg.$$

Three milligrams per kilogram is the maximum amount or Irganox that can migrate from the bag into the food product. The specific migration limit for this substance is 6 mg/kg (according to Directive 2002/72), which means that in this particular case, given the initial concentration, the legal limit of migration is not achieved. Alternatively, the actual migration value can be measured either in food simulants or in the real food products.

Example 2 Butadiene monomer from chewing gum polymer (Leber, 2001)

The exposure to butadiene (B) monomer from chewing gum polymer styrene butadiene rubber (SBR), considering worse-case assumptions (1) "heavy user" of chewing gum (3 g/day), 2.4% of styrene butadiene rubber polymer in the gum, and (3) the residual amount of butadiene monomer in the styrene butadiene rubber is 500 ppb, is given by

$$\text{Average exposure} = 3\,g/day \times 2.4\% \times 0.5\,\mu g\,B\,/\,g\,SBR = 36\,ng\,B/day,$$

Presently in the EU the specific migration limits are set assuming that for the substance under consideration:

- Every person eats 1 kg/day, over a lifetime, packaged in the material that contains that substance.
- The food (1 kg) is in contact with 6 dm^2 of packaging material.
- Average consumer body weight is 60 kg.
- There is no other significant source of exposure.

The specific migration limit is set as the maximum value that yields an intake lower than the TDI or other safe criteria applicable. This set of assumptions may be used as worst-case assumptions together with actual average migration data to point estimate the intake.

Example 3 Bisphenol A diglycidyl ether intake from canned fish (Simoneau et al., 1999)

The exposure to bisphenol A diglycidyl ether (BADGE) from canned fish in oil is assessed from the per capita consumption of canned fish and the average concentration of BADGE found in surveyed samples:

Consumption (kg/person/year) = annual production tonnage/population = 31,000 tons/10 million = 3.1 kg/person/year.

BADGE average concentration found in fish = 0.14 mg/kg.
Average exposure = 0.43 mg BADGE/person/year=1.17 µg BADGE/person/day.

An average weight of 60 kg for an adult is assumed-

Average exposure = 0.003 µg BADGE/kg body weight/day.

7.5   Probabilistic Approach for Modeling Exposure Assessment

In contrast to the deterministic models for exposure that use a single estimate of each variable, probabilistic models take account of every possible value that each variable can take and weight each possible scenario by the probability of its occurrence. This approach ensures that any variability and/or uncertainty in variables, including food consumption, are reflected in the model output. Variability (true heterogeneity) refers to temporal, spatial, or interindividual differences in the value of an input. Uncertainty (lack of knowledge) may be thought of as a measure of the incompleteness of one's knowledge or information about an unknown quantity whose true value could be established if a perfect measuring device were available. Random and systematic errors are sources of uncertainty (Cullen and Frey, 1999).

Food consumption data are confronted with uncertainty and variability. The observed values differ from the true value by systematic errors or bias (which occurs, on average, in the measurements of all measured subjects), and nonsystematic errors that vary unpredictably from subject to subject within the population under study (Lambe, 2002). Migration data are also subject to variability owing to heterogeneity in the packaging system and in the composition and structure of the food product, and uncertainty regarding, for example, the distribution chain (e.g., temperature). At the laboratory scale, variability in migration value determination is also unavoidable, even when a standard procedure is used.

Probabilistic modeling using Monte Carlo or Latin hypercube simulations is an empirical method to determine variability and/or uncertainty and a procedure that can examine a number of factors using individual distributions for the model variables, such as food consumption, migration levels of a substance, and body weight of the consumer. Single-point data are drawn from each distribution repeatedly, looking at numerous combinations of input data to give a frequency distribution of exposure. Some examples using probabilistic modeling are briefly outlined in Sect. 7.5.2 after some words on the selection of the function to represent the distribution of the data for the model variables.

7.5.1 Data and Distributions

Both variability and uncertainty may be quantified using probability distributions and a very important part of any probabilistic exposure assessment to food chemicals is the selection of probability distributions for the uncertain input variables (Lambe, 2002).

A probability distribution model is typically represented mathematically in the form of either a probability density function or a cumulative distribution function. Table 7.3 shows the probability density function of some of the more common probability distribution models.

The cumulative distribution function is obtained by integrating the probability density function (Fig. 7.1). This provides a relationship between percentiles – percentage of values that are less than or equal to a given value – and quartiles (value of the random variable associated with a given percentile). In exposure assessments, is important to know what portion of a population faces an exposure less than or equal to some level, such as the 90th or 95th percentile.

These parametric distributions include lognormal, normal, beta, Weibull, and gamma among others. In the area of food contaminants, the lognormal distribution is often assumed (Gilsenan et al., 2003). The distributions may or may not be truncated: Normal distributions should be truncated at both ends and lognormal distributions should be truncated at the high end to avoid unrealistic values. Body weight, for example, is often described as a normal or lognormal distribution, while triangular distributions have been used to represent uncertainties affecting parameters estimated by expert judgment. Concentration data often approximate a lognormal distribution.

The key characteristics of a distribution are the central tendency, the dispersion and the shape. These descriptive parameters summarize information about the distribution (Table 7.4).

Table 7.3 Selected probability distribution models used in exposure assessment

Model	Some characteristics	Probability distribution function
Lognormal	• Only nonnegative values • Often used to represent large asymmetric uncertainties • Tail-heavy; may not be the best if a good fit to the upper tail is required • Widely used to represent concentration data • For example., human body weights	$$f(x) = \frac{1}{\sqrt{2\pi}\sigma x}\exp\left(-\frac{\left(\ln(x) - \ln m\right)^2}{2\sigma^2}\right)$$ for $0 \leq x \leq \infty$, where m is the median or scale parameter and σ is the standard deviation of $\ln x$ or shape parameter
Normal	• It is not a default that can be assumed to apply unless proven otherwise • Has infinite tails • The coefficient of variation should be less than 0.3 to represent adequately nonnegative quantities • For example, human heights	$$f(x) = \frac{1}{\sqrt{2\pi}\sigma}\exp\left(-\frac{\left(x - \mu\right)^2}{2\sigma^2}\right)$$ for $-\infty \leq x \leq \infty$, where μ is the mean and σ is the standard deviation
Exponential	• Useful for representing the time interval between successive random, independent events that occur at a constant rate	$$f(x) = \lambda\exp(-\lambda x)$$ for $0 \leq x \leq \infty$
Beta	• Very flexible • With a finite upper and lower bound (commonly 0,1) • Used to represent judgments about uncertainty • Useful in Bayesian statistic	$$f(x) = \frac{x^{\alpha 1 - 1}\left(1 - x\right)^{\alpha 2 - 1}}{B\left(\alpha 1, \alpha 2\right)}$$ for $0 \leq x \leq 1$, and $\alpha 1, \alpha 2 > 0$ $$B\left(\alpha 1, \alpha 2\right) = \frac{\Gamma\left(\alpha 1\right)\Gamma\left(\alpha 2\right)}{\Gamma\left(\alpha 1 + \alpha 2\right)}$$ $$\Gamma\left(x\right) = \left(x - 1\right)!$$ $\alpha 1$, $\alpha 2$ are the shape parameters
Uniform	• Special case of beta • Used frequently in exposure assessment • Useful to represent subjective judgments about uncertainty when an expert is willing (able to estimate an upper and lower bound of a quantity)	$$f(x) = \frac{1}{b - a}$$ for $a \leq x \leq b$, $$\text{mean} = \left(a + b\right)/2,$$ $$\text{variance} = \left(b - a\right)^2/12$$
Triangular	• Maximum entropy distribution used to represent variability and uncertainty when only upper and lower bounds and a most likely value are known	$$f(x) = \frac{b - \left|x - a\right|}{b^2}$$ for $a - b \leq x \leq a + b$

(continued)

Table 7.3 (continued)

Model	Some characteristics	Probability distribution function
Weibull	• Useful for representing processes such as the time to completion or the time to failure • Can assume negatively, symmetric, or positively skewed shape • May also be used to represent nonnegative physical quantities • Is less tail-heavy than lognormal • May provide a better fit to the upper percentiles of a particular dataset • Estimation of parameters involves the solution of nonlinear equations	$f(x) = \dfrac{\alpha}{\beta}\left(\dfrac{x-L}{\beta}\right)^{\alpha-L} \exp\left[-\left(\dfrac{x-L}{\beta}\right)^{\alpha}\right]$ for $x \geq L, \alpha, \quad \beta > 0$ α is the shape parameter, β is the scale parameter and L represents the location

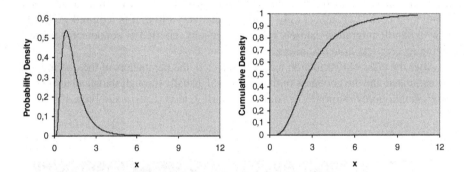

Fig. 7.1 Lognormal distribution for median $m = 1$ and standard deviation $\alpha = 0.6$

Table 7.4 Summary of important characteristics and description of parameters of distributions

Key characteristics	Parameters
Central tendency	Mean $- \mu_1$ (first moment of a distribution), value for which the weighted probability for all values lower than the mean is equal to the weighted probability for all values higher than the mean Median – corresponds to the 50th percentile of the distribution Mode – value associated with a maximum of the probability density function
Dispersion	Variance $- \mu_2 = \sigma^2$ (second central moment of the distribution with respect to the mean) Standard deviation – square root of the variance
Shape	Skewness $- \gamma_1 = \mu_3/\sigma^3$ represents the asymmetry of a distribution (based on the third moment) Kurtosis $- \gamma_2 = \mu_4/\sigma^4$ represents the peakedness of a distribution (based on the fourth moment)

The selection of a distribution model to represent a given dataset is very important to the output of the probabilistic analysis (Liptom et al., 1995). A first step to follow is the probability plot, which can give an idea of the overall shape of the distribution. Some characteristics of the variable are useful for identifying classes of distributions (Parmar et al., 1997):

- Whether the variable is continuous or discrete
- The bounds of the values
- Symmetry of the values
- Range of values of more concern

The relationship between the square of the skewness (β_1) and the kurtosis (β_2) can also be very useful for the distribution selection. Figure 7.2 shows this relationship for a few commonly used distributions. For example, the normal distribution always has a skewness of 0 and a kurtosis of 3 and, therefore, is represented as a single point.

Depending on the objective or purpose of the study, it must be defined whether the distributions of food consumption should reflect the distribution for the total population or for consumers only. Additionally, very often, it is required to decide how to handle migration datasets with many results reported as nondetected values, owing to very low concentrations.

After the class of distribution has been selected, the parameters of the distribution are estimated and the goodness of fit is evaluated, usually through statistical tests such as χ^2, Kolmogorov–Smirnov, or Anderson–Darling tests (Cullen and Frey, 1999).

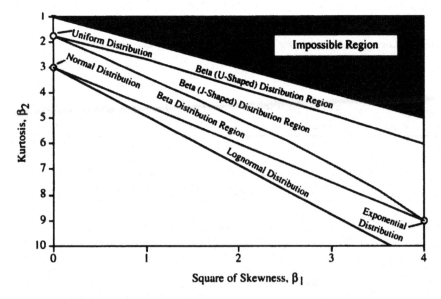

Fig. 7.2 Plane β_1–β_2 for several common parametric probability distributions (Cullen and Frey, 1999)

7.5.2   Probabilistic Modeling

There are several ways to propagate information about variability or uncertainty through a model (Cullen and Frey, 1999). This means, in the general model as described by Eq. 7.1, that there are several ways of combining migration information and food consumption information that are described by distribution functions. There are analytical and approximation methods, and numerical simulation methods, such as Monte Carlo and Latin hypercube sampling (Petersen, 2000).

The Monte Carlo simulation is most often used in exposure assessments. In this simulation method, a model is run repeatedly, using different values for each of the uncertain input parameters each time. The values of each of the uncertain input parameters are generated on the basis of the probability distribution for the parameter. The model is then executed as it would be for any deterministic analysis. The process is repeated until the specified number of model iterations has been completed. The result, then, is a set of values for each of the model output variables, in the form of a distribution of calculated results, which can be treated statistically as if they were an experimentally or empirically observed set of data (Cullen and Frey, 1999). Using Monte Carlo simulation, one can, therefore, represent uncertainty in the output of a model. As an example, Fig. 7.3a shows a cumulative exposure distribution to a given substance for a consumer population. The population members consume different quantities of different food items, which are in different types of packages, and therefore are exposed to different levels of the substance. The reading off from the graphs shows, for example, that 95% of all population is exposed to 4.5 µg/day or less of the substance. Then this value can be compared with the TDI or any other safety level of consumption. The results are affected by uncertainty in the measurements and sampling of the migration data, in the ratios between packaging contact area and the amount of food (as actual packages show ratio values different from the conventionally assumed 6 dm^2/kg), and market share. Figure 7.3b shows the results output to a given substance showing the 95% confidence range of each consumer percentile.

The sample size corresponds to the number of repetitions used in the numerical simulation. The selection of sample size may follow different approaches: to decide on an acceptable confidence interval for the mean, developing a confidence interval

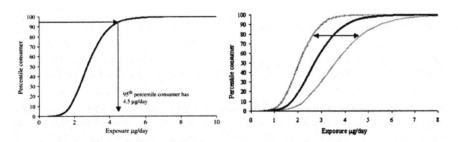

Fig. 7.3   Example of output results from Monte Carlo simulation (Holmes et al., 2005)

for whatever fractile level is of most concern in the investigation, and considering the degree of accuracy required.

The Monte Carlo simulation may yield important insights into the sensitivity of the model to variations in the input parameters, as well as into the likelihood of obtaining any particular outcome. Monte Carlo methods also allow for the use of any type of probability distribution for which values can be generated using a computer, rather than restricting one to forms which are analytically tractable.

It is possible to simulate jointly distributed random variables in which correlations may exist and assess the impact on the model output distribution of including or neglecting correlations among the inputs.

Probabilistic modeling has been employed to assess the exposure to chemicals with origin in packaging materials. The model developed by Central Science Laboratory in the UK was applied to estimate the short-term exposure of UK consumers to epoxidized soybean oil (ESBO), a plasticizer from metal closures used in glass jars, and to residual BADGE, from can coatings of canned foods. The model quantifies variability and uncertainty and uses a two-dimensional Monte Carlo simulation which enables the separation of the influence of variability and uncertainty in the outputs. The model is supported by UK consumption data from surveys and quantifies variability by estimating exposure for each population member using concentrations sampled from distributions based on measurements in real foods or food simulants (inner loop). Uncertainty is quantified in an outer loop, by repeating the inner loop with different assumptions about model inputs that are uncertain (consumption and concentrations measurements, sampling, extrapolation of migration data between simulants and real foods, and proportion of food packaged) (Holmes et al., 2005). As an example, the results of the inner loop are illustrated in Fig. 7.3a (where the exposure for the different days for each individual are averaged) and the result from the outer loop is illustrated in Fig. 7.3b.

Outputs include estimated exposures for specified percentiles of the population (e.g., median and 97.5th percentile consumers) with confidence limits. Sensitivity analysis shows which sources of variability and uncertainty have the most influence on the exposure estimates. The examples 4 and 5 are of outputs from this model.

7.6   Conclusions

Probabilistic modeling has proved to be an excellent tool in a number of research areas, including exposure assessment of food additives, pesticides, and other contaminants. Its application to exposure assessment of substances migrating from packaging is regarded as of great interest when both food consumption data and migration data are of good quality. However, it is recognized that adequate guidelines must be drawn up for the performance of exposure assessments on migrants from food contact materials, the comparison of currently used European Food Safety Authority procedures with stochastic modeling, and particularly the use of the Monte Carlo technique.

Example 4 ESBO from metal lids in glass jars (SEFEL, 2004)

Data:
UK consumption data extrapolated to EU
Actual surveyed body weights
Assumptions:
Case A
– All foods that could be packaged in jars are considered to be packaged in jars
– Fifty percent of soups assumed to be packaged in jars
– All these foods contain the maximum concentration of ESBO found: 530 mg/kg
Case B
– The same as in case A but with actual migration data from samples collected from the market

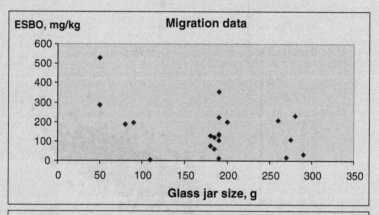

Results:

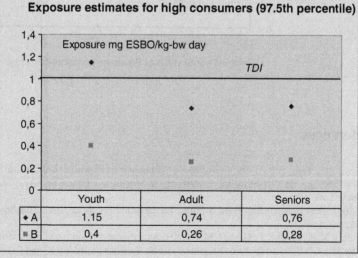

	Youth	Adult	Seniors
◆ A	1.15	0,74	0,76
▪ B	0,4	0,26	0,28

Example 5 BADGE from canned food (Holmes et al., 2005)

Data:
UK consumption data extrapolated to EU
Actual surveyed body weights
Food Standards Agency studies (BADGE concentration in canned foods)

Results:

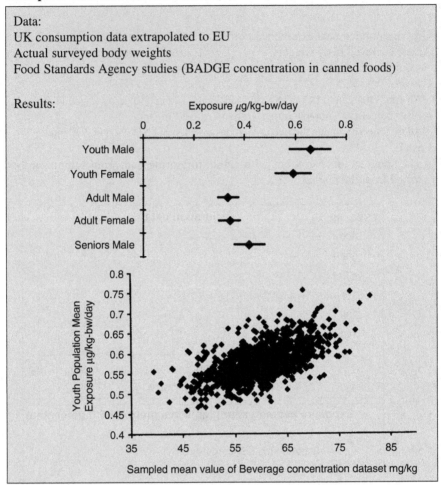

References

Cullen, A.C.; Frey, H.C., 1999. Probabilistic Techniques in Exposure Assessment: A Handbook for Dealing with Variability and Uncertainty in Models and Inputs. Society of Risk Analysis, New York

EC, 2001. Report on Dietary Food Additive Intake in the European Union. Health & Consumer Protection DG

Gilsenan, M.B.; Lambe, J.; Gibney, M.J., 2003. Assessment of food intake input distributions for use in probabilistic exposure assessments of food additives. *Additives and Contaminants*, 20 (11), 1023–1033.

Hart, A.; Smith, G.C.; Macarthur, R.; Rose, M., 2003. Application of uncertainty analysis in assessing dietary exposure. Toxicology Letters, 140–141, 437–442.

Holmes, M.J.; Hart, A.; Northing, P.; Oldring, P.K.T.; Castle, L., 2005. Dietary exposure to chemical migrants from food contact materials: a probabilistic approach. *Food Additives and Contaminants*, 22 (10), 907–919.

ILSI, 2001. Report Series – Exposure from Food Contact Materials – Summary Report of a Workshop held in October 2001, ILSI Europe

Kroes, R.; Muller, D.; Lambe, J.; Lowik, M.R.H.; Klaveren, J.; Kleiner, J.; Massey, R.; Mayer, S.; Urieta, I.; Verger, P.; Visconti, A., 2002. Assessment of intake from the diet. *Food and Chemical Toxicology*, 40, 327–385.

Lambe, J., 2002. The use of food consumption data in assessments of exposure to food chemicals including the application of probabilistic modelling. Proceedings of the Symposium "Nutritional Aspects of Food Safety". *Proceedings of the Nutrition Society*, 61, 11–18.

Leber, A.P., 2001. Human exposures to monomers resulting from consumer contact with polymers. Chemico-Biological Interactions, 135–136, 215–220.

Lipton, J.; Shaw, W.D.; Holmes, J.; Patterson, A., 1995. Selecting input distributions for use in Monte Carlo simulations. *Regulatory Toxicology and Pharmacology*, 21, 192–198.

Luetzow, M., 2003. Harmonization of exposure assessment for food chemicals: the international perspective. Toxicology Letters, 140–141, 419–425.

Parmar, B.; Miller, P.F.; Burt, R., 1997. Stepwise approaches for estimating the intakes of Chemicals in food. *Regulatory Toxicology and Pharmacology*, 26, 44–51.

Petersen, B.J., 2000. Probabilistic modelling: theory and practice. Food Additives and Contaminants, 17 (7), 591–599.

Poças, M.F.; Hogg, T., 2007. Exposure assessment of chemicals from packaging materials in foods: a review. *Trends in Food Science & Technology*, 18, 219–230

Rees, N.M.A.; Tennant, D.R., 1993. Estimating consumer intakes of food chemical contaminants. In: Watson, D.H. (Ed.), Safety of Chemicals in Food: Chemical ContaminxHorwood, Chichester, pp. 175–181.

SEFEL, 2004. Exposure assessment of ESBO from jar lids using the CSL stochastic model with industry input on package and food selections. SEFEL-Food Contact Commission FCC223/04

Simoneau, C.; Theobald, A.; Wiltschko, D.; Anklam, E., 1999. Estimation of intake of BADGE from canned fish consumption in Europe and migration survey. *Food Additives and Contaminants*, 16 (11), 457–463.

WHO, 1997. Guidelines for Predicting Dietary Intake of Pesticide Residues. Document WHO/FSF/FOS/97.7. World Health Organization, Geneva.

Vose, D., 2000. Risk Analysis: A Quantitative Guide. Wiley, West Sussex.

Chapter 8
Migration from Packaging Materials

B. De Meulenaer

8.1 Introduction

Various chemical compounds can be present in foodstuffs which may induce health problems in humans. The origin of these compounds can be very diverse. Mathematical modeling can sometimes be used to predict the concentration of these chemicals in the food. Particularly for compounds which are produced in the food during, e.g., processing and for compounds which migrate from a food contact material this technique can be very fruitful. For the former type of compounds, classical chemical kinetics can be applied. In this contribution, the modeling of the migration from polymeric food contact materials is considered. This migration phenomenon can be modeled mathematically since the physical processes which govern this process are very well studied and understood. Therefore, initially some of these fundamentals will be discussed in more detail.

8.2 Physical Processes Which Govern the Migration Process

As mentioned in Chap. 1, migration from food contact materials should be considered as a submicroscopic mass transfer from the polymeric food contact material to the food. Three subprocesses should be distinguished in the whole migration process. First of all, a low molecular weight compound will diffuse in the polymer in the direction of the food owing to the presence of a concentration gradient (diffusion process). Subsequently, reaching the food–plastic interface, the migrant will be desorbed by the polymer and absorbed by the food (sorption process). Finally the migrant, currently dissolved in the food, will diffuse into the total food matrix, again owing to the presence of a concentration gradient. Alternatively, the latter

B. De Meulenaer, (✉)
Department of Food Safety and Food Quality, Ghent University, Coupure Links 653,
9000 Ghent, Belgium

R. Costa, K. Kristbergsson (eds.), *Predictive Modeling and Risk Assessment,*
DOI: 10.1007/978-1-387-68776-6, © Springer Science+Business Media, LLC 2009

mass transfer can be accelerated by a convection process inside the food matrix as will be discussed in more detail later.

So, in conclusion, the migration process is governed by diffusion and sorption. The sorption and diffusion process can be described quantitatively by using the partition coefficient $K_{P/F}$ and the diffusion coefficients D_P and D_F, where the indexes P and F refer to the polymer or plastic and the food respectively.

8.2.1 Partition Coefficient and the Sorption Process

The partition coefficient of component a between a polymer and a food can be defined as follows:

$$K_{P/F(a)} = \frac{C_{P,a,\infty}}{C_{F,a,\infty}},\qquad(8.1)$$

where $C_{P,a,\infty}$ and $C_{F,a,\infty}$ are, respectively, the equilibrium concentration of the component a in the polymer and the food (Franz, 2000). Basically, this definition is derived from the assumption that in equilibrium conditions ($t=\infty$) the chemical potential of the migrating substance in the polymer, μ_{P_a}, is equal to the chemical potential of the same compound in the food, μ_{F_a} (Baner, 2000).

Mainly the partition coefficient depends on the polarity of the substance and of the polarity of the two phases involved. The following simple example illustrates the importance of the partition coefficient.

At equilibrium conditions, the amount substance a that has migrated into the food, $m_{F,a,\infty}$, can be calculated as follows. Supposing that initially no migrating substance was present in the food ($m_{F,a,0} = 0$) and that $m_{P,a,0}$ represents the initial amount of migrant present in the polymer, then it can be concluded from the mass balance that

$$m_{P,a,0} = m_{F,a,\infty} + m_{P,a,\infty}.\qquad(8.2)$$

From Eq. 8.1, however,

$$m_{P,a,\infty} = \frac{m_{F,a,\infty}}{V_F} \times K_{P/F(a)} \times V_P,\qquad(8.3)$$

where V_P and V_F are, respectively, the volume of the polymer and the volume of the food.

Consequently,

$$m_{F,a,\infty} = \frac{m_{P,a,\infty}}{1 + K_{P/F(a)} \times \dfrac{V_P}{V_F}}.\qquad(8.4)$$

Generally it can be assumed that $V_p/V_F \ll 1$. From Eq. 8.4 it can be concluded that in the case $K_{P/F(a)} \ll 1$, the amount of migrated substance in the food in equilibrium conditions ($m_{F,a,\infty}$) equals the initial amount of substance present in the polymer ($m_{P,a,0}$), which implies that total migration of the substance out of the polymer occurred. If, in contrast, a hydrophobic substance is applied in a hydrophobic polymer contacted with water, it is clear that $K_{P/F(a)} \gg 1$ and from Eq. 8.4, it can be concluded that

$$m_{F,a,\infty} \ll m_{P,a,\infty}, \tag{8.5}$$

indicating that migration remains restricted. Since most polymers used are (fairly) hydrophobic and most of the low molecular weight compounds present in a plastic are hydrophobic as well, it follows from the above example that migration will be especially important in apolar food matrices, like fatty foods. Of course, polar migrants, such as the antistatic poly(ethylene glycol), will preferentially migrate to more polar, so called aqueous foods.

Partition coefficients can be determined experimentally, but this approach often is very tedious and prone to experimental errors. Therefore, empirical methods were developed to estimate the partition coefficients for given polymer–migrant–food systems. Detailed discussion of these methods falls outside the scope of this chapter.

8.2.2   Diffusion Coefficient and the Diffusion Process

The diffusion coefficient D_a of compound a in a particular matrix follows from Fick's first law, stating that the mass flux of a compound a in the x direction, $J_{x,a}$, during time t through a unit area is proportional to the gradient of the concentration of the compound, C_a, considered in the x direction. Mathematically this gives the following:

$$J_{x,a} = -D_a \frac{\partial C_a}{\partial x}. \tag{8.6a}$$

Of course, fluxes in the other directions can be defined similarly

$$J_{y,a} = -D_a \frac{\partial C_a}{\partial y} \tag{8.6b}$$

and

$$J_{z,a} = -D_a \frac{\partial C_a}{\partial z}. \tag{8.6c}$$

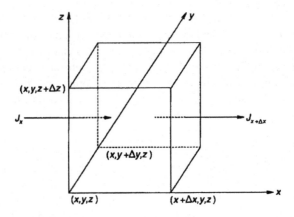

Fig. 8.1 Diffusion through an elementary volume

Owing to this flux, however, the concentration in a unit cell of the matrix in which the diffusion is taking place, with dimensions $\Delta x \Delta y \Delta z$, will vary accordingly as a function of time (Fig. 8.1).

The net change of the concentration of compound a in this unit cell as a function of time can be found from

$$-\frac{\partial C_a}{\partial t} = \frac{[J_{x,a}(x+\Delta x) - J_{x,a}(x)]\Delta y \Delta z}{\Delta x \Delta y \Delta z}$$
$$+\frac{[J_{y,a}(y+\Delta y) - J_{y,a}(y)]\Delta x \Delta z}{\Delta x \Delta y \Delta z} + \frac{[J_{z,a}(z+\Delta z) - J_{z,a}(z)]\Delta x \Delta y}{\Delta x \Delta y \Delta z}. \quad (8.7)$$

If the unit cell becomes infinitely small, then Eq. 8.7 becomes

$$-\frac{\partial C_a}{\partial t} = \frac{\partial J_{x,a}}{\partial x} + \frac{\partial J_{y,a}}{\partial y} + \frac{\partial J_{z,a}}{\partial z} \qquad (8.8a)$$

or if the diffusion coefficient is constant:

$$-\frac{\partial C_a}{\partial t} = D_a\left(\frac{\partial^2 C_a}{\partial x^2} + \frac{\partial^2 C_a}{\partial y^2} + \frac{\partial^2 C_a}{\partial z^2}\right). \qquad (8.8b)$$

Equation 8.8a is known as the general Fick's second law of diffusion.

Various models which are able to predict the diffusion coefficient for a given thermoplastic polymer–substance system have been developed throughout the years. It is not the intention to review those models in detail in this chapter. It should be emphasized, however, that the diffusion mechanisms in thermoplasts below and above the glass-transition temperature of the polymer are totally different. Basically three different cases can be considered depending on the diffusion rate of the migrant and the relaxation rate of the polymer (the relaxation rate of the polymer is related to the time the polymer needs to adjust itself to a new equilibrium). In case 1, diffusion or Fickian diffusion occurs when the diffusion rate is much less than the relaxation rate of the polymer; in case 2, diffusion is characterized by a rapid diffusion in comparison with the relaxation of the polymer; and in case 3, anomalous diffusion occurs when both the diffusion and the relaxation rates are comparable (Schlotter and Furlan, 1992).

Rubbery polymers adjust quickly to changes in their physical condition, so consequently the diffusion of low molecular weight compounds is considered to be Fickian. When the temperature is reduced to below the glass-transition temperature of the polymer, the molecular mobility of the polymer will be highly reduced and diffusion corresponds mostly to case 2 or case 3 in these circumstances. Therefore, estimation of diffusion coefficients in glassy polymers is even more difficult than for rubbery polymers.

In most cases, the various models used are able to predict the diffusion coefficient of relatively small molecules such as gases or water. Currently, however, estimation of the diffusion coefficient of a migrant in a given polymer system is not possible. Therefore, empiric models have been used to estimate the diffusion coefficient of various migrants in particular polymeric systems, as will be discussed later. These models, however, are based on dynamic migration studies. To elucidate the relationship between the experimentally observed migration dynamics and the diffusion coefficient, the migration phenomenon should be approached from a mathematical approach.

8.3 Mathematical Approach to Estimate Migration from Plastics

8.3.1 General Transport Equation

The goal of the mathematical models describing migration from plastic food contact materials is to predict the concentration of the migrant in the food after contact with the plastic. In such a manner, lengthy and costly migration experiments which are legally required can be avoided. In addition, from this mathematical approach, relevant parameters controlling the migration process can be identified. Knowledge of these parameters is of prime importance to allow realistic simulations of the migration phenomenon in the laboratory and to interpret the results obtained in the correct manner.

A reliable model should consider all mass-transfer phenomena and other processes affecting the concentration of the migrant in the food. Basically the following processes are taken into account:

- Diffusion of the migrant
- Convection of the medium in which the migrant is dissolved
- Chemical reactions in which the migrant is involved

It is important to realize that from a theoretical point of view all these processes can take place in both the food and the polymer. Practically, however, mainly the following processes control the migration behavior:

- Diffusion of the migrant in both the polymer and the food
- Chemical reaction in both the polymer and the food

Convection of the polymer is very much restricted in normal conditions of use, so will have no influence with regard to the migration. In liquid foods, convection will cause a quick distribution of the migrant in the food favoring a uniform concentration of the migrant. In solid foods or highly viscous foods, diffusion of the migrant will be of higher importance than convection fluxes of the food itself.

The chemical reactions a migrant is subjected to are also an interesting feature. An antioxidant present in the polymer undergoes partial degradation during the production of the plastic contact material, lowering the potent migration of the compound considered. In the food itself, migrants may undergo reactions as well. Bisphenol A diglycidyl ether, a cross-linking agent used in epoxy coatings for food cans, is an important example.

Mathematically, the predominant processes affecting the concentration C of the migrant at a particular place with coordinates (x,y,z) in the food–polymer system can be written as follows:

- For diffusion

$$-\frac{\partial C}{\partial t} = \frac{\partial J_x}{\partial x} + \frac{\partial J_y}{\partial y} + \frac{\partial J_z}{\partial z}. \tag{8.8c}$$

- For the chemical reaction

$$\frac{\partial C}{\partial t} = -kC^n. \tag{8.9}$$

Equation 8.8c represents the previously introduced second diffusion law of Fick and in Eq. 8.9, n represents the order of the chemical reaction and k is the reaction rate constant.

Combining the two gives the general transport equation in which the convection in both the polymer and the food is supposed to have a minor influence:

$$\frac{\partial C}{\partial t} = \left(\frac{\partial^2 DC}{\partial x^2} + \frac{\partial^2 DC}{\partial y^2} + \frac{\partial^2 DC}{\partial z^2} \right) - kC^n. \tag{8.10}$$

Of course the main problem in solving this equation lies in the second-order partial differential equation. Such an equation has only an analytical solution in some special cases. In addition, the diffusion coefficient should be constant. In all other cases, numerical methods should be used to solve the equation. Once this equation has been solved however, the problem of a reliable estimate of the diffusion coefficient of the migrant remains. As discussed previously, mechanistic and atomistic diffusion models are currently unable to give reliable predictions for the diffusion coefficient of migrants in plastics.

The second part in this equation is rather specific for particular migrants and will not be discussed here in more detail; therefore, only solutions to the second-order differential equation will be presented.

As indicated, a number of assumptions should be made to analytically solve the second-order partial differential equation given in Eq. 8.10. Primarily, the diffusion coefficient is supposed to be constant in both the food and the polymer. In addition, it is assumed that diffusion takes place in only one direction, perpendicular to the surface of the polymer. Consequently, the partial differential equation becomes

$$\frac{\partial C}{\partial t} = D \frac{\partial^2 C}{\partial x^2}. \tag{8.11}$$

Solutions to this equation for finite and infinite polymers will be presented in the following sections.

8.3.2 Diffusion from Finite Polymers

Further basic assumptions in addition to those mentioned already include the following:

- There is one single migrant, which is uniformly distributed in the polymer at $t = 0$ at concentration $C_{P,0}$.
- The concentration of the migrant in the food at a particular time, $C_{F,t}$, is everywhere the same (which implies that the food is ideally mixed).
- A constant distribution of the migrant between the polymer and the food takes place according to

$$K_{P/F} = \frac{C_{P,t}}{C_{F,t}} = \frac{C_{P,\infty}}{C_{F,\infty}}. \tag{8.12}$$

- The contact material is a flat sheet.
- The mass transfer is mainly controlled by diffusion taking place in the polymer.

The following solution to Eq. 8.11 was developed for a polymer in contact with a finite food (Crank, 1975; Piringer, 2000; Hamdani et al., 1997):

$$\frac{m_{F,t}}{m_{F,\infty}} = 1 - \sum_{n=0}^{\infty} \frac{2\alpha(1+\alpha)}{1+\alpha+\alpha^2 q_n^2} \exp\left(-\frac{D_P q_n^2 t}{L_P^2}\right), \tag{8.13}$$

in which $m_{F,t}$ is the amount of migrant in the food at a particular time t, q_n is the positive root of the trigonometric identity $\mathrm{tg}(q_n) = -\alpha q_n$, L_P is the thickness of the polymer, D_P is the diffusion coefficient of the migrant in the polymer, and α is given by the following formula:

$$\alpha = \frac{1}{K_{P/F}} \frac{V_F}{V_P} = \frac{m_{F,\infty}}{m_{P,\infty}}, \tag{8.14}$$

in which V_F and V_P represent the volume of, respectively, the food and the polymer and $m_{P,\infty}$ is the mass of the migrant present in the polymer at equilibrium conditions ($t = \infty$).

The rather complex Eq. 8.13 can be simplified by assuming the finite polymer is contacted with an infinite food. This implies that the concentration of the migrant in the food is zero, since mathematically speaking $V_F \infty$. Consequently, from Eq. 8.14 it follows that $\alpha \gg 1$. Equation 8.13 can be simplified to (Piringer, 2000; Hamdani et al., 1997)

$$\frac{m_{F,t}}{m_{F,\infty}} = 1 - \sum_{n=0}^{\infty} \frac{8}{(2n+1)^2 \pi^2} \exp\left(-\frac{(2n+1)^2 \pi^2}{4L_P^2} D_P t\right). \tag{8.15}$$

Equation 8.15 is reported to give the same results as Eq. 8.13 if the volume of the food (V_F) exceeds 20 times the volume of the polymer (V_P), which in practice is usually achieved, also in migration tests (Hamdani et al., 1997).

Equation 8.15 can further be simplified for the following two cases (Hamdani et al., 1997):

1. Long contact time ($m_{F,t}/m_{F,\infty} > 0.6$)

$$\frac{m_{F,t}}{m_{F,\infty}} = 1 - \frac{8}{\pi^2} \exp\left(-\frac{\pi^2}{L_P^2} D_P t\right) \tag{8.16}$$

2. Short contact time ($m_{F,t}/m_{F,\infty} < 0.6$)

$$\frac{m_{F,t}}{m_{F,\infty}} = \frac{2}{L_P} \sqrt{\frac{D_P t}{\pi}} \tag{8.17}$$

For all these models, it was assumed that diffusion in the polymer is the main factor controlling the migration phenomenon. If other processes, such as the dissolution and the diffusion of the migrant in the food, are also important factors to consider, analytical solutions of the diffusion equation (Eq. 8.11) are not available. Numerical methods for some cases have been described (Laoubi and Vergnaud, 1996).

A further simplification of the problem, assuming the polymer is infinite, allows one in some cases to take into account the dissolution and the diffusion of the migrant in the food as explained in the following section.

8.3.3   Diffusion from Infinite Polymers

The assumption of an infinite polymer implies that the concentration of the migrant in the polymer is constant as a function of time ($C_{P,0} = C_{P,t}$). Of course, this does not correspond to reality since it is known that the concentration of the migrant in the polymer is affected by migration (Hamdani et al., 1997). Again, several solutions of Eq. 8.11 have been proposed for a number of cases taking into account the following supplementary boundary conditions:

- There is one single migrant, which is uniformly distributed in the polymer at $t = 0$ at concentration $C_{P,0}$.
- A constant distribution between the polymer and the food takes place according to Eq. 8.12.
- The contact material is a flat sheet.

Two major cases can be distinguished depending on the concentration gradient of the migrant in the food.

8.3.3.1   No Concentration Gradient of the Migrant in the Food

If no concentration gradient in the food is present, this implies that the food is well mixed or that the diffusion of the migrant in the food proceeds much faster than the diffusion in the polymer. The general solution of Eq. 8.11 is given by Limm and Hollifield (1995), Piringer (2000), Lickly et al. (1997), and Hamdani et al. (1997):

$$m_{F,t} = \frac{C_{P,0}A}{K_{P/F}}\left(1 - e^{z^2} \text{etfc}(z)\right) \tag{8.18}$$

in which A is the contact surface between the polymer and the food and z is given by

$$z = \frac{K_{P/F}\sqrt{D_P t}}{A}. \tag{8.19}$$

This equation is valid for infinite polymers contacted with finite foods, indicating that the migrant slowly dissolves in the food "a" (Hamdani et al., 1997). Consequently, migration is mainly governed by solvation.

If the migrant is very soluble in the food however ($K_{P/F} \ll 1$), Eq. 8.18 can be simplified to (Piringer, 1994; Lickly et al., 1997)

$$\frac{m_{F,t}}{A} = 2C_{P,0}\sqrt{\frac{D_P t}{\pi}}. \tag{8.20}$$

Equation 8.20 represents the migration from an infinite polymer in contact with an infinite food ($C_F = 0$) (Hamdani, 1997). In this case, diffusion of the migrant in the polymer will dominate the migration process.

8.3.3.2   Concentration Gradient of the Migrant in the Food

If a concentration gradient of the migrant is present in the food, the following equation has been proposed as a solution to Eq. 8.11 (Piringer, 2000):

$$\frac{m_{F,t}}{A} = 2C_{P,0}\sqrt{\frac{D_p t}{\pi}}\left(\frac{\beta}{\beta+1}\right),$$

(8.21)

in which

$$\beta = \frac{1}{K_{P/F}}\sqrt{\frac{D_F}{D_P}}.$$

(8.22)

As can be noticed, diffusion of the migrant in the food and in the polymer are taken into account. If in this case diffusion in the food is high ($\beta \gg 1$), Eq. 8.21 reduces to Eq. 8.20, indicating that owing to the high diffusion in the food, the concentration gradient of the migrant in the food is negligible.

If on the other hand, $\beta \ll 1$, because of the poor solubility of the migrant in the food, migration will be especially dominated by the migration in the food as indicated in the following equation derived from (8.20) (Piringer, 2000):

$$\frac{m_{F,t}}{A} = \frac{2C_{P,0}}{K_{P/F}}\sqrt{\frac{D_F t}{\pi}}.$$

(8.23)

8.4   Estimation of Material Constants

As can be concluded from all the analytical solutions to the general diffusion equation (Eq. 8.11), diffusion coefficients and the partition coefficient of the migrant should be known to practically apply these equations. As will be explained later, from a regulatory point of view, the "worst-case" scenario predicting migration is of primary interest. Therefore, it is most frequently assumed that the solubility of the migrant in the polymer is very high, which implies that $K_{P/F} = 1$, thus avoiding difficulties for the estimation of the partition coefficient for a given migrant–polymer–food system.

The main problem of a realistic estimate of the diffusion coefficient remains. Diffusion coefficients of migrants range from about 10^{-7} cm^2/s down to about 10^{-18} cm^2/s. From the previous equations it can be concluded that this large difference in magnitude will play a major role in the final migration result for most cases. Realistic estimates are therefore considered indispensable because underestimated diffusion coefficients will underestimate migration and overestimates will make the practical use of these migration models impossible (Brandsch et al., 2000).

The mechanistic models on diffusion currently available however can not be applied for the estimation of diffusion coefficients of migrants in polymers. Alternatively, empirical formulas such as Eqs. 8.24 and 8.25 can be used:

$$D_P = D_0 exp\left(\xi\sqrt{M_r} - \psi\frac{\sqrt[3]{M_r}}{T}\right),$$

(8.24)

in which D_0, ξ, and ψ are constants related to the activation energy of diffusion, the molecular weight of the migrant, T is the absolute temperature, and M_r is the molecular weight of the migrant considered (Limm and Hollifield, 1996) and

$$D_P = 10^4 exp\left(A_P - \frac{\tau}{T} - 0.1351M_r^{2/3} + 0.003M_r - \frac{10,454}{T} \right), \qquad (8.25)$$

in which A_p is a so-called polymer-specific diffusion conductance parameter and τ is a polymer-specific activation energy parameter.

These empiric equations are based on the linkage of a migration-dynamics database and the general migration model previously presented (Eq. 8.13). Therefore, it can be stated that no real independent validation is made, since the diffusion coefficients obtained, using these empiric equations, are used again in the migration models, from which the diffusion coefficients were derived. Although from a scientific point of view, this is a serious drawback, Eqs. 8.24 and 8.25 are believed to be very important from a practical point of view, since currently these are the only means to obtain diffusion coefficients in a simple manner.

Reliable diffusion coefficients for migrants having a molecular weight up to 4,000 could be calculated in such a way for selected polyolefins between the melting temperature and the glass-transition temperature of the polymer (Brandsch et al., 2000).

For non-polyolefins however, which are characterized by a higher glass-transition temperature (frequently between 50 and 100°C), such models are not currently available owing to lack of experimental data. Therefore, no useful diffusion coefficient estimates are available for these polymers.

8.5 Practical Use of Mathematical Models

There is a general consensus about the usefulness of mathematical modeling of the migration phenomenon to limit laboratory tests which are tedious and costly. This is reflected by the possibility to use mathematical modeling in order to prove compliance with legislation as recently accepted within the EU and as has been accepted before in the USA. In addition, extensive information on the use of these models is available in the *Practical Guide*, issued by the European Commission. In this document, reference is made to two software packages which are available on the market (MIGRATEST Lite 2001, FABES, Munich, Germany) or can be freely downloaded from the Internet (SMEWISE, INRA, Reims, France). Both programs basically use the same migration model and the same empirical equation to determine the diffusion coefficient of a particular polymer–migrant system. The numerical methods to solve Eq. 8.13 may be different. Nevertheless, both programs are reliable if they are properly used by trained persons, with a more than basic knowledge of the migration phenomenon.

Mathematical models, however, are prone to a number of limitations which are important to consider. As indicated before, an important aspect in the evaluation of the models is the correspondence between the calculated and the experimental data. Because of the necessity of a reliable estimate of the diffusion coefficient of the migrant, it can be concluded from the above discussions that currently only models applicable to polyolefins are available. It should be noted, however, that polyolefins are currently the most frequently applied polymers for food contact. On the other hand, it should be realized that in the migration models applied, it is supposed that the diffusion coefficient of the migrant is constant. If fatty foods are contacted with polyolefins, however, negative migration of the triacylglycerols will occur, resulting in a time-dependent change of the diffusion coefficient of the migrants in the polymeric system. Since it was concluded previously that especially migration to fatty foods is of concern, this fact can be considered as a major disadvantage.

The models described are only able to predict the migration of known and well-characterized migrants. Consequently the models will not be able to predict the total amount of substance that migrated from a contact material, since the contact material may contain apart from the additives also a number of other compounds of which the identity is not completely known (e.g., ethylene oligomers or their breakdown products present in polyethylene).

Care should be taken in using too simplified models such as the one discussed above. The use of a more general equation such as Eq. 8.13 is considered to be better. Specific migration of various additives from polyolefins to olive oil at various time–temperature conditions with the predicted values of this migration model are compared in Fig. 8.2. For polypropylene, almost all the estimated values are higher

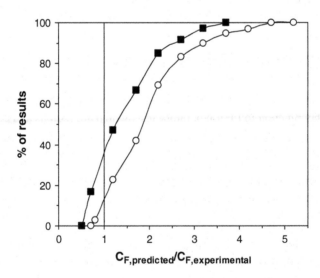

Fig. 8.2 Comparison between the predicted and experimental migration levels in olive oil for several additives out of high density polyethylene (*filled square*) and polypropylene (*open circle*). (Based on the experimental data reported in O'Brien et al. 1999, 2001)

than those experimentally observed. From a safety point of view this is interesting. Results for polyethylene were a bit less promising. To obtain a more accurate estimation, it was believed that the use of a more realistic $K_{P/F}$, especially at lower temperatures, would be appropriate.

References

Baner, A.L. (2000) Partition coefficients, in "Plastic packaging materials for food, barrier function, mass transport, quality assurance and legislation", O.-G.Piringer and A.L.Baner (Eds.), Wiley-VCH, Weinheim, pp. 79–123.

Brandsch, J., Mercea, P. and Piringer, O. (2000) Possibilities and limitations of migration modelling, in "Plastic packaging materials for food, barrier function, mass transport, quality assurance and legislation", O.-G.Piringer and A.L.Baner (Eds.), Wiley-VCH, Weinheim, pp. 446–468.

Crank, J. (1975) "Mathematics of diffusion", Clarendon, London, 212 p.

Food contact materials: practical guide. (2003) European Commission Health & Consumer Protection Directorate-General. Downloaded from http://ec.europa.eu/food/food/chemical-safety/foodcontact/practical_guide_en.pdf on 26 September 2008.

Franz, R. (2000) Migration of plastic constituents, in "Plastic packaging materials for food, barrier function, mass transport, quality assurance and legislation", O.-G.Piringer and A.L.Baner (Eds.), Wiley-VCH, Weinheim, pp. 287–357.

Hamdani, M., Feigenbaum, A. and Vergnaud, J.M. (1997) Prediction of worst case migration from packaging to food using mathematical models. *Food Additives and Contaminants*, 14, 49–506.

Laoubi, S. and Vergnaud, M. (1996) Theoretical treatment of pollutant transfer from a polymer packaging made of recycled film and a functional barrier. *Food Additives and Contaminants*, 13, 293–306.

Lickly, T.D., Rainey, M.L., Burgert, L.C., Breder, C.V. and Borodinsky, L. (1997) Using a simple diffusion model to predict residual monomer migration – considerations and limitations. *Food additives and Contaminants*, 14, 65–74.

Limm, W. and Hollifield, H. (1995) Effect of temperature and mixing on polymer adjuvant migration to corn oil and water. *Food Additives and Contaminants*, 12, 609–624.

Limm, W. and Hollifield, H. (1996) Modelling of additive diffusion in polyolefins. *Food Additives and Contaminants*, 13, 949–967.

O'Brien, A., Goodson, A. and Cooper, I. (1999) Polymer additive migration to foods – a direct comparison of experimental data and values calculated from migration models for high density polyethylene (HDPE). *Food Additives and Contaminants*, 16, 167–380.

O'Brien, A., Cooper, I. and Tice, P.A. (2001) Polymer additive migration to foods – a direct comparison of experimental data and values calculated from migration models for polypropylene. *Food Additives and Contaminants*, 18, 343–355.

Piringer, O. (1994) Evaluation of plastics for food packaging. *Food Additives and Contaminants*, 11, 221–230.

Piringer, O. (2000) Transport equations and their solutions, in "Plastic packaging materials for food, barrier function, mass transport, quality assurance and legislation", O.-G.Piringer and A.L.Baner (Eds.), Wiley-VCH, Weinheim, 183–219.

Schlotter, N.E. and Furlan, P.Y. (1992) A review of small molecule diffusion in polyofefins. *Polymer*, 33, 3323–3342.Resources

De Meulenaer, B. and Huyghebaert, A. (2005) Migration residues from food contact materials, in "Handbook of Food Analysis", L.Nollet (Ed.), Marcel Dekker, New York, ISBN 0-8247-5073-3, pp. 1297–1330.

Part III
Environment and Raw Food Production

Chapter 9
Antibiotics in Animal Products

Amílcar C. Falcão

9.1   Introduction

The administration of antibiotics to animals to prevent or treat diseases led us to be concerned about the impact of these antibiotics on human health. In fact, animal products could be a potential vehicle to transfer drugs to humans. Using appropriated mathematical and statistical models, one can predict the kinetic profile of drugs and their metabolites and, consequently, develop preventive procedures regarding drug transmission (i.e., determination of appropriate withdrawal periods). Nevertheless, in the present chapter the mathematical and statistical concepts for data interpretation are strictly given to allow understanding of some basic pharmacokinetic principles and to illustrate the determination of withdrawal periods.

9.2   Antibiotics in Animals

9.2.1   Introduction

The antibiotics discussed here are drugs currently used in veterinary medicine to destroy or inhibit the growth of bacteria. Most of them are naturally produced by bacteria and fungi, while others are man-made but have the same target. The introduction of antibiotics has saved countless lives. Without antibiotics, pet and farm animals would endure pain and suffering and the safe production of food would be endangered.

Like human medicines, by law all veterinary medicinal products require official authorization before they can be marketed. For this reason, full dossiers on quality, effectiveness, and safety for animals and especially safety for users, consumers, and

A. C. Falcão (✉)
University of Coimbra, Faculty of Pharmacy, Rua do Norte, 3000-295, Coimbra, Portugal

R. Costa, K. Kristbergsson (eds.), *Predictive Modeling and Risk Assessment,*
DOI: 10.1007/978-1-387-68776-6, © Springer Science + Business Media, LLC 2009

the environment should be submitted to the authorities. Of special importance is the evaluation and operation of withdrawal periods, which are the times during which treated animals cannot be slaughtered for food, nor can their products enter the human food chain (Anon, 2004).

9.2.2   Use of Antibiotics in Animals

The appropriate use of antibiotics has given man and animals freedom from many diseases and safe and efficient food production on farms. Nevertheless, it is important to be aware that antibiotics cover a very wide range of compounds, split into many different classes with individualized pharmacotoxicological characteristics.

There is currently concern that use of antibiotics in animals may compromise the effectiveness of related medicines in man. However, it is recognized that antibiotics are essential for animal health and welfare. As a result, extensive studies, set up in conjunction with the relevant regulatory authorities, are under way to help shed more light on this subject (WHO, 1997).

The use of antibiotics in animal medicine has three main objectives: to treat an individual or an outbreak of disease; to prevent outbreaks occurring; and to enable farm animals to derive optimum benefit from their food (Anon, 2004).

9.2.2.1   Treatment of Diseases

As also happens with humans, when animals fall ill, they may need medicines, and courses of antibiotics are given under prescription whenever necessary. Antibiotics may be given in different ways, such as by tablet, by injection, in drinking water, or in food and their prescription will be for as short a period as necessary to beat the disease.

Most farm animals are relatively young and, like children, are susceptible to many infectious diseases. Without antibiotics, if they contract a bacterial disease their condition can deteriorate rapidly, resulting in suffering and, potentially, death. Disease can also spread to other animals and in some cases humans; therefore, where livestock are grouped together, it is often necessary to treat the whole group even when only a few are showing signs of illness to prevent others from being infected.

9.2.2.2   Prevention of Diseases

Prevention is always better than cure, particularly in species where disease can spread rapidly. To prevent outbreaks of infectious bacterial disease when animals are most at risk, and animals are known to be susceptible, courses of appropriate antibiotics may be prescribed by a veterinary surgeon (anticoccidial products are a good example for prevention and control of a disease).

9.2.2.3   Production Purposes

A selected group of antibiotics, reserved for animal use only, is commonly prescribed to help growing animals digest their food more efficiently, get maximum benefit from it, and allow them to develop into strong and healthy individuals. This can be achieved by destroying or inhibiting undesirable bacteria in the gut which promote the optimum absorption of food.

Where individual dosing of animals becomes impractical, the only systems that are suitable for large groups are drinking-water medication or in-feed medication (the latter is still the most widely used for technical reasons).

9.3   Residues of Antibiotics in Animal Products

The presence of residues of drugs and feed additives in animal products is one the main concerns of the authorities. To protect the health of the consumer of foodstuffs of animal origin, one the most important principles laid down in the legislation is that foodstuffs obtained from animals treated with veterinary medicinal products must not contain residues of the medicine or its metabolites which might constitute a health hazard for the consumer. To facilitate the uniform application of this principle throughout the European Union, and to ensure that differences in the assessment of the effects of residues by member states do not create barriers to the free movement of foodstuffs of animal origin, on June 26, 1990 the Council adopted Regulation (EEC) No. 2377/90 laying down a procedure for the establishment of maximum residue limits (MRLs) for veterinary medicinal products in foodstuffs of animal origin European Commission (2003). This regulation was subsequently amended to introduce values of MRLs for substances used in veterinary medicinal products in the annexes to the regulation and to adapt the legislation in light of technical progress and amendments to the regulatory procedures (Anadón and Martínez-Larrañaga, 1999).

With the adoption of Regulation No. 1308/99, the legislation made the European Agency for the Evaluation of Medical Products (EMEA) established in 1995 responsible for the procedure to process applications for the establishment of MRLs. The EMEA is responsible for coordinating the existing scientific resources placed as its disposal by the competent authorities of the member states for the evaluation and supervision of medicinal products. The EMEA provides scientific opinion on applications for centralized marketing authorization for both human and veterinary use as well as the establishment of MRLs. The EMEA comprises the Committee for Proprietary Medicinal Products (CPMP) for human products and the Committee for Veterinary Medicinal Products (CVMP). The CVMP is responsible for preparing the opinion of the EMEA on any question relating to evaluation of veterinary medicinal products and a secretariat provides technical administrative support and ensures appropriate co-ordination (European Commission, 2003).

9.3.1 Definition of MRLs

The concept of residues of veterinary medicinal products means all pharmacologically active substances, whether active principles, excipients or degradation products, and their metabolites which may be remained in foodstuffs obtained from animals to which the veterinary medicinal product in question has been administered.

In Council Regulation (EEC) No. 2377/90, a MRL is defined as "the maximum concentration of residue resulting from the use of a veterinary medicinal product (expressed in mg kg^{-1} or g kg^{-1} on a fresh weight basis) which may be accepted by the Union to be legally permitted or recognized as acceptable in or on a food." It should be noted that this definition is virtually the same as that adopted by the FAO/WHO Codex Alimentarius Committee for Residues of Veterinary Drugs in Foods.

The approach used by the CVMP (EMEA) for the evaluation of the safety of residues is similar to the approach used by other committees and international scientific bodies charged with the safety evaluation of food additives and contaminants, based on the determination of a no-effect level and the use of safety factors to determine an acceptable daily intake (ADI) on which subsequently MRLs are based.

Under European Union legislation, all pharmacological substances used in food-producing animals must be entered into one of three annexes of Council Regulation (EEC) No. 2377/90. These are as follows: Annex I, full MRLs; Annex II, no MRLs required on consumer safety grounds; Annex III, provisional MRLs (pending further data). Annex IV is the destination of drugs considered unsafe on consumer health grounds. Drugs in the Annex IV are effectively prohibited for use in food-producing animals within the European Union (Anadón, 1990).

The EMEA publishes the document *Status of MRL Procedures* (http://www.emea.eu.int/pdfs/vet/srwp/076599en.pdf), where all substances included in Annex I, Annex II, Annex III, or Annex IV of Council Regulation (EEC) No. 2377/90 are listed in alphabetical order with the indication of the respective annex and the number of the regulation which established these MRLs. In addition, the list includes substances for which MRL recommendations have been made by the CVMP and are now undergoing the decision-making procedure by the European Commission. The list also includes the substances for which the CVMP concluded that owing to insufficient data provided a recommendation for inclusion of the substance in Annex I, Annex II, or Annex III of Council Regulation (EEC) No. 2377/90 could not be made.

9.3.2 The ADI Concept

After the completion of the various pharmacological, toxicological, and other tests undertaken to demonstrate the safety of the substance, the first major stage in the process is the establishment of the ADI. The ADI is an estimate of the drug residues (parent and all metabolites), expressed in terms of micrograms or milligrams per kilogram of body weight, that can be ingested daily over a lifetime without any appreciable health risk to exposed individuals.

The basis for the calculation of the ADI is the no-observed-(adverse)-effect level with respect to the most sensitive parameter in the most sensitive appropriate test species, or in some cases, in humans. A safety factor is then applied to provide a margin of safety, taking into account the inherent uncertainties in extrapolating animal toxicity data to human beings and to take account of variations within the human species. In selecting a safety factor, one usually assumes that human beings may be up to 10 times more sensitive than the test animal species and that the difference in sensitivity within the human population is a tenfold range. Thus, where good-quality data are available, a safety factor of 100 is usually applied, although this may be increased or reduced depending on the nature and the quality of the data available and the effects observed in animals or humans (WHO, 1989).

Once the ADI has been agreed upon, it is then necessary to determine MRLs for the individual food commodities concerned. Since the ADI is related to body weight, an arbitrary average human body weight is defined as 60 kg. The ADI expressed on a microgram or milligram per kilogram of body weight basis is therefore multiplied by 60 to give the total amount of residue which may be ingested by an individual. In addition, consideration also has to be given to the actual levels of consumption of foods of animal origin. Since accurate consumption figures are difficult to obtain, and there are in any case substantial variations between individual consumers, arbitrarily high fixed values are used to ensure protection of the majority of consumers. Thus, for example, to derive MRLs from the ADI it is assumed that the average person consumes, on a daily basis, 500 g of meat 1.5 l of milk, and 100 g of eggs or egg products. The total amount of residues present in this daily food basket is not allowed to exceed the ADI (European Commission, 2003).

9.3.3 Antimicrobial Drugs

The food chain represents the major route of transmission of resistant bacteria and/ or resistance genes from animals to humans and the exposure of animal gut flora to antimicrobial veterinary medicinal products is expected to play a major role in such a possible transmission. Therefore, the potential public health risk of antimicrobial resistance as a result of the use of antimicrobial veterinary medicinal products in animals depends not only on the importance of the antimicrobials for human medicine but also on the level of exposure of the animal gut flora to the antimicrobial veterinary medicinal product (Anon, 2003).

In fact, when we are talking about the properties of the active substances used in veterinary medicine, account must be taken not only of the toxicological properties of the substances in the limited sense of the term (such as teratogenic, mutagenic, or carcinogenic effects) but also of their pharmacological properties and their possible immunotoxic potential. Moreover, in the case of antibiotics and similar substances, the possibility of a microbiological risk including the development of antimicrobial drug resistant bacteria in the human gut flora may also need to be considered. It must be emphasized that the impact of low levels of antibiotics on the intestinal microflora is not directly examined using classic toxicology studies. For substances with microbiological activity what is actually used is the major adverse microbiological effect arising from the effects of residues of antimicrobial drugs in products of animal origin acting on the human gastrointestinal bacteria flora, particularly those acting on the colonic flora (European Commission, 2003).

9.4   Clinical Pharmacokinetics

The specialized study of the mathematical relationships between a drug dose regimen and resulting drug concentrations (usually in serum or plasma) is known as the study of pharmacokinetics. It is a quantitative science, where drug absorption, distribution, and elimination are defined by numerical values. Stated in simpler terms, pharmacokinetics can be considered as "what the body does to the drug" (Birkett, 1998).

The application of pharmacokinetic principles to the safe and effective therapeutic management of animals is known as clinical pharmacokinetics. With the advent of clinical pharmacokinetics, practitioners have acquired a tool for individualizing dose regimens to accommodate interpatient variability in dose/systemic drug concentration/drug responses. In relation to drug residues, special attention is required to the distribution of a drug in the animal's body, its metabolic fate, and the elimination of the parent compounds and its metabolites. Kinetic data have been used to establish dosing intervals and total body elimination of the active compounds. Elimination half-life can be used to predict the magnitude of drug accumulation with multiple dosing, the time needed to reach the steady sate, and the time needed to reduce blood concentrations to some target concentration. Nevertheless, drug kinetic changes may occur among blood and tissues, so tissue depletion studies are needed to establish the withdrawal times which ensure residue levels below MRLs (Anadón and Martínez-Larrañaga, 1992).

9.4.1   The Physiologic Approach

The pharmacokinetic parameters of bioavailability, volume of distribution, and body clearance are determined by physiologic processes in the body. The following descriptions of drug absorption, drug distribution, and hepatic and renal clearance serve to show how physiologic and pharmacokinetic processes are interrelated.

9.4.1.1 Absorption and Bioavailability

Drugs are administered by either intravascular (intravenous) or extravascular (oral, intramuscular, etc.) routes. A variety of routes and dose formulations of drugs are available. The absorption, by definition, is the movement of a drug from the site of administration to the blood circulation. The bioavailability of a particular dose form includes the assessment of both the rate of drug absorption and the extent of drug absorption from the site of administration, such as the gastrointestinal tract, skin, or muscle. However, it must be emphasized that the bioavailable fraction of a drug and the time course of the drug concentration following administration are influenced not only by the intrinsic characteristics of the selected dose form but also by the route of administration, and the physiologic status of the animal (Roland and Tozer, 1995).

9.4.1.2 Distribution

Once the drug enters the bloodstream, it undergoes simultaneous distribution to body tissues and elimination by clearing organs (liver and kidney). Only drug that is not bound to protein is free to be distributed to extravascular tissues in a reversible way. The pharmacokinetic term most often used to characterize the distribution of a drug is the apparent volume of distribution (V_d). This volume does not necessarily correspond to a physiologic space and, therefore, is preceded by the word "apparent." The apparent volume of distribution is the hypothetical volume that would be required to account for all drug in the body, according to

$$V_d = \text{amount of drug in the body / serum drug concentration} \qquad (9.1)$$

Once distribution of the drug to tissues has been completed and equilibrium has been established, the amount of drug in the body can be estimated from the product of serum concentration and the apparent volume of distribution. The apparent volume of distribution is also used to estimate the amount of drug in the body when the serum concentration is known, and to project the approximate increase in serum concentration following the rapid absorption or injection of a drug dose (Birkett, 1998).

9.4.1.3 Elimination

Elimination of most drugs usually occurs by a first-order process in which the rate of drug elimination is directly proportional to the serum concentration. As the serum drug concentration rises, more of the drug is available to the elimination organ, and subsequently the drug will be eliminated at a faster rate. Because of the linear relationship between rate of elimination and concentration, a first-order process is also known as a linear process. The pharmacokinetic term "clearance" (CL) best describes the efficiency of the elimination process. Clearance by an eliminating organ is the volume of serum, plasma, or blood that is totally cleared of drug per unit time. This term is additive; the total body (systemic) clearance of a drug is equal to the sum of the clearances by individual eliminating organs. Usually this is presented as the sum of hepatic (metabolism) and renal (excretion) clearances:

$$CL = CL_H + CL_R, \tag{9.2}$$

where CL_H is hepatic clearance and CL_R is renal clearance. Clearance is constant and independent of serum concentration for drugs eliminated by first-order processes, and therefore may be considered a proportionality constant between the rate of drug elimination and the serum concentration (Brikett, 1998):

$$CL = \text{rate of elimination} / \text{serum drug concentration.} \tag{9.3}$$

In addition to clearance, another pharmacokinetic parameter may be used to describe elimination under certain conditions. Half-life $(t_{1/2})$ is the time required for the serum concentration to decrease by 50% (Fig. 9.1). After one half-life, 50% of the drug in the body has been eliminated, 75% after two half-lives, and 87.5% after three half-lives, etc. (Table 9.1).

The half-life is mathematically derived from the elimination rate constant (k_e), which has units of inverse time and may be thought of as the fractional rate of drug elimination from the body. If $k_e = 0.25$ h^{-1}, then 25% of the amount of drug in the

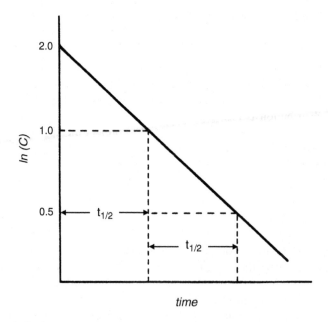

Fig. 9.1 The half-life concept. The concentration falls by half each half-life

Table 9.1 Fraction of drug lost and fraction remaining in the body after each half-life

Number of half-lives	Fraction lost	Fraction remaining
1	0.500	0.500
2	0.750	0.250
3	0.875	0.125
4	0.938	0.062
5	0.969	0.031
6	0.984	0.016

body will be eliminated per hour at any given moment in time, recalling that the amount of drug in the body is always decreasing with time during elimination. The elimination rate constant relates the change in serum concentration with time (dC/dt) with serum concentration (C) according to a first-order process:

$$dC / dt = -k_e \times C, \tag{9.4}$$

where the negative sign indicates elimination. On integration of this expression, one obtains

$$C_2 = C_1 \times e^{-k_e (t_2 - t_1)}, \tag{9.5}$$

where C and t represent serum concentrations and their corresponding times. To illustrate the relationship between the elimination rate constant and half-life, if the quantity $(t_2 - t_1)$ represents one half-life, then $C_2/C_1 = 0.5$ and

$$0.5 = e^{-(k_e \times t_{1/2})} \tag{9.6}$$

or

$$t_{1/2} = 0.693 / k_e. \tag{9.7}$$

Like clearance, the elimination rate constant is independent of the serum concentration under first-order conditions. This is because the elimination rate constant is determined in part by clearance. The elimination rate constant is also influenced by the volume of distribution according to

$$k_e = CL / V_d. \tag{9.8}$$

Although the elimination rate constant and elimination half-life are terms usually associated with elimination (i.e., clearance), it must be noted that they are also influenced by drug distribution. It is important to understand this concept when individualizing drug therapy, since an increase of the apparent volume of distribution may result in a prolonged half-life without any change in drug elimination efficiency (Shargel and Yu, 1999).

9.4.2   Pharmacokinetic Modeling

9.4.2.1   Purpose of Modeling

To characterize the absorption, distribution, and elimination of a drug, studies are performed to determine pharmacokinetic parameters that will allow for appropriate dose prediction. The serum concentration versus time data obtained from individuals are fit to various pharmacokinetic models, and statistical methods are used to determine the best fit. These calculated pharmacokinetic parameters are used to predict dose regimens in other settings. However, caution should always be exercised when applying population data to an individual, particularly when physiopathologic conditions and/or potential drug interactions exist in the individual that were not present in the study group (Gabrielsson and Weiner, 2000).

9.4.2.2   One-Compartment Model

When a drug is given in the form of a rapid intravenous injection (intravenous bolus), the entire dose of drug enters the bloodstream immediately, and the drug absorption process is considered to be instantaneous. In addition, the one-compartment model with first-order elimination assumes that the drug is distributed immediately from the vascular space to tissues with instantaneous attainment of equilibrium. In fact, the one-compartment model offers the simplest way to describe the process of drug distribution and elimination in the body in accordance with the following model (Fig. 9.2) and the corresponding mathematical approach:

$$C_t = \frac{D}{V_d} e^{k_e t}, \tag{9.9}$$

where Ct is the concentration at time t, D is the dose administered, V_d is the apparent volume of distribution, and k_e is the elimination rate constant.

According to this model, a plot of the logarithm of serum concentration (y-axis) versus time (x-axis) following an intravenous injection yields a straight line representing elimination from a single compartment. The same data on a rectilinear plot show a curvilinear line (Fig. 9.3).

Most drugs given orally may be adequately described using a one-compartment model, whereas drugs administered by rapid intravenous infusion are usually best described by a two-compartment model (Roland and Tozer, 1995).

9.4.2.3   Two-Compartment Model

Many drugs given in a single intravenous bolus dose demonstrate a plasma level–time curve that does not decline as a single exponential (first-order) process as happens with the one-compartment model. A drug that follows the pharmacokinetics of a

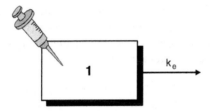

Fig. 9.2 One-compartment model for a drug administered by rapid intravenous injection

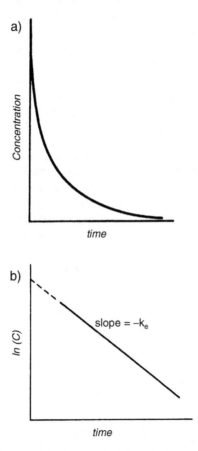

Fig. 9.3 Concentration–time courses of a drug that displays one-compartment kinetics, plotted linearly (**a**) and semilogarithmically (**b**)

two-compartmental model does not rapidly equilibrate throughout the body as is assumed for a one-compartment model. In this model, the drug is distributed into two compartments, the central compartment (blood, extracellular fluid, and highly perfused tissues) and the peripheral compartment (tissues in which the drug equilibrates more slowly). By convention, compartment 1 is the central compartment and compartment 2 is the tissue compartment. Drug transfer between the two compartments is assumed to take place by first-order processes (Fig. 9.4).

Mathematically, the two-compartment model after a single intravenous injection (intravenous bolus) can be expressed by the following equation:

$$C_t = \frac{D}{V_1}\frac{\alpha - k_{21}}{(\alpha - \beta)}e^{-\alpha t} + \frac{D}{V_1}\frac{k_{21} - \beta}{(\alpha - \beta)}e^{-\beta t}, \tag{9.10}$$

where C_t is the concentration at time t, D is the dose administered, V_1 is the apparent volume of distribution for the central compartment, k_{21} is the first-order rate transfer constants for the movement of drug from compartment 2 to compartment 1, and α and β are macroconstants (or hybrid constants) that depend solely on k_{21}, k_{12} (the inverse of k_{21}), and k_{10} (the first-order rate constant for the movement of the drug from compartment 1 to the exterior). Equation 9.10 can be simplified into the following expression:

$$C_t = Ae^{-\alpha t} + Be^{-\beta t}, \tag{9.11}$$

where the constants A and B are intercepts on the y-axis for each exponential segment of the curve. Therefore, a biexponential (biphasic) logarithmic serum concentration versus time curve is usually observed when a drug is administered intravenously and the plasma level–time curve for a drug that follows a two-compartment model may be divided into two parts, a distribution (α) phase and an (β) elimination phase. Their corresponding half-lives are estimated as

$$t_{1/2\alpha} = 0.693 / \alpha \tag{9.12}$$

and

$$t_{1/2\beta} = 0.693 / \beta \tag{9.13}$$

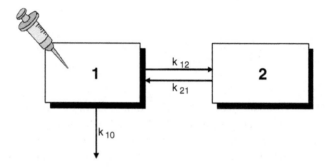

Fig. 9.4 Two-compartment model for a drug administered by rapid intravenous injection

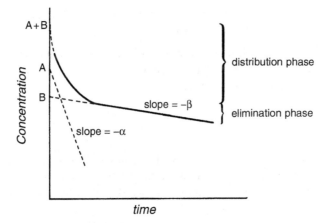

Fig. 9.5 Semilogarithmic plot of the biexponential decay in concentration of a drug administered as an intravenous bolus dose

The initial, rapid decline (α phase) of serum concentrations is due to distribution of the drug from the central compartment to the tissue compartment. The duration of this distribution phase may be estimated by a time equal to 5 times the distribution half-life. The decline in concentration after that time (β phase) will primarily reflect elimination in a two-compartment pharmacokinetic model (Fig. 9.5). In practice, the β phase is almost equivalent to the decline of serum concentrations observed when the one-compartment model is adopted (Roland and Tozer, 1995).

9.4.3   Clinical Pharmacokinetics in Residue Studies

In accordance with the *Notice to Applicants and Note for Guidance* (European Commission, 2003) regarding consumer safety, the aim of the pharmacokinetic studies is to evaluate the absorption, distribution, metabolism, and excretion of the product in the target species. Data should demonstrate the time course of the concentrations of the parent drug and/or its metabolite(s) in tissue and body fluids. The study is normally carried out in healthy animals. If diseased animals are used, this should be indicated.

9.4.3.1   Absorption

The absorption of the substance concerned should be documented. Depending on the route of administration of any substance, the following parameters should be studied, if applicable:

- Following oral administration, absorption of animal species, differences in digestive physiologic function, and sites of metabolism (e.g., first-pass effects)
- Absorption following topical application
- Release from injection sites

- Systemic availability following other specific routes (intramammary application, intrauterine administration, etc.)

In the case of substances, if it is proven that systemic absorption is negligible, then further residue studies are not required. However, if there is significant systemic absorption, full residue studies are required.

9.4.3.2 Distribution

The distribution should be described by pharmacokinetic parameters and specific influences depending on the animal physiologic function, plasma protein binding, and accumulation after repeated administrations should be discussed.

9.4.3.3 Metabolism

Different biotransformation products may possess different toxic potentials; therefore, information on the chemical nature, the concentration, and the persistence of the total residue is required. The purpose of the metabolism study in the target species of an animal is to provide the necessary information on the metabolic fate of the drug in the edible tissues. These studies are also necessary to check whether the metabolite(s) found in target animals are the same as those found in laboratory animals used for toxicity testing.

All the available data from any application in the target species, the test species (including in vitro systems), and – as far as available – data from human use should be considered. The studies should provide the following types of information, if necessary for an adequate evaluation of the metabolism of the product:

- Nature of the metabolite(s)
- Chemical nature of residues in edible tissues (muscle, liver, kidney, fat) and in products (milk, eggs and honey)
- Influence on the metabolic profile of the route of administration
- Relations between structure and biological activities, if available
- Bioavailability of bound metabolite(s), characterization of the type of binding of biologically relevant drug metabolite(s) to constituents of edible tissues

9.4.3.4 Excretion

Data on the excretion in target species is requested: renal excretion (always) and fecal excretion (if necessary). All other routes of excretion should be considered when appropriate (milk, expired air, etc.).

9.4.3.5 Depletion Residues

The measure of depletion of total drug-related residue in edible tissues of target animals after the last administration of the drug must be carried out. The radiotracer method is currently the most useful technique for determining the total drug-related residue. However, pending establishment of the ADI, alternatives may be used, such as a microbiological method to measure the total microbiologically active residues.

The results of residue depletion studies should indicate whether and, if so, for how long after the application of the veterinary medicinal product residues occur in the foodstuffs obtained from treated target animals.

9.5   Withdrawal Period: A Rational Approach

One of the clear trends that has emerged in recent years is the concept of risk analysis in regulatory assessment (Anon, 2001), which includes several steps in the overall process: (1) risk assessment which comprises hazard identification, hazard characteristics, exposure assessment, and risk characteristics; (2) risk management; and (3) risk communication.

The idea of zero risk, which seemed to be the goal of some authorities in regulatory assessment, is now recognized as unrealistic. The harmonization of the establishment of the MRL throughout the European Union gives a good illustration of how risk analysis can be applied. Risk management in this context means setting appropriate withdrawal periods. Risk communication means making the information available in a fully transparent manner, and explaining the extent of the risk and how it is managed in a way that puts it into overall perspective (Jones, 1997).

9.5.1   The Withdrawal Period: Concept

The withdrawal period is defined in Directive 2001/82/EC as "the period necessary between the last administration of the veterinary medicinal product to animals under normal conditions of use and the production of foodstuff from such animals to ensure that such foodstuffs do not contain residues in quantities in excess of the maximum residue limits laid down in application of Regulation (EEC) No 2377/90."

The withdrawal period should provide a high degree of assurance both to the producers and to the consumers that the concentrations of residues in foods derived from treated animals are not above the permitted concentrations. Thus, MRLs are the key point as can be seen by the CVMP guidance on the setting of withdrawal periods (Anon, 1996).

In accordance with guidelines, MRLs should be established for each edible tissue of each target species for which the pharmacologically active substance is intended. Irrespective of the species to which the active substance is administered, there is substantial agreement that the MRL should, where possible, be the same in each species as the hazard characterization of the residue is essentially similar and several safety factors have been used in its derivation. Considering the knowledge on the variation of residue depletion within classes of animals and therefore on the exposure assessment, the risk characterization should also not differ substantially within an animal class. Therefore, an extrapolation of the MRL from one species to further species within a class of animals is considered as the default approach (Table 9.2).

In addition, as general rule, where residue concentrations can be quantified in the edible tissues MRLs should be established for all edible tissues on the basis of the tissue residue distribution pattern of the pharmacologically active substance investigated (Table 9.3).

9.5.2 Withdrawal Period: Background

In accordance with the approach towards harmonization of the determination of withdrawal periods (Anon, 1996), the "statistical method" must be applied for *new* chemical entities, on which the residue depletion studies should be appropriately conducted and data will be sufficiently adequate to use the procedure mentioned. In the case of *old* chemical entities, whenever available data are insufficient to evaluate the withdrawal period by a statistical method, the use of a "simplified method" can be adopted (but is not recommended because it is unacceptable in terms of statistical inference): the withdrawal period corresponds to the time point at which the concentrations of residues in all tissues for all animals fall below the respective MRLs. However, when one has determined that time point, the estimation of a safety span should be considered to compensate for the uncertainties of biological variability. Although an overall recommendation cannot be provided, an approximate guide for a safety span is likely to be a value of 10–30% of the time period, when all observations are below the MRL. Alternatively, a safety span might be calculated as well from the tissue depletion half-life, possibly a value of

Table 9.2 Extrapolation of maximum residue limits (*MRLs*)

Species for which MRLs have been set	Extrapolations to
Major ruminant	All ruminants
Major ruminant milk	All ruminants' milk
Major monogastric mammal	Extrapolation to all monogastric mammals
Chicken and eggs	Poultry and poultry eggs
Salmonidae	All fin fish
Either a major ruminant or a major monogastric mammal	Horses

Table 9.3 Edible tissues and products

Mammals	Poultry	Fish	Bees
Muscle	Muscle	Muscle and skin in natural proportions	Honey
Liver	Liver		
Kidney	Kidney		
Fat, or fat and skin in natural proportions (pigs)	Fat and skin in natural proportions		
Milk	Eggs		

1–3 times $t_{1/2}$. A minimal safety span (e.g., 1–2 days) should be considered in any case (Vanic et al., 2003).

9.5.3   Withdrawal Period: Determination

Although the "statistical method" is strongly recommended by the CVMP for determination of the withdrawal period, the adopted statistical approach can be selected, justified, and supported with adequate documentation by the applicant. In fact, parametric (EMEA and FDA suggestions) and nonparametric approaches are available (Vanic et al., 2003). Nevertheless, in the present context, the "statistical method" described next is the "linear regression" technique recommended by the CVMP and explained in detail in the EMEA/CVMP/036/95/FINAL document (Anon, 1996).

9.5.3.1   Linear Regression

The depletion curve is modeled as a straight line after a logarithmic transformation. The terminal elimination of a drug (the depletion residue) in most cases follows a one-compartmental model and is sufficiently described by one exponential term. The first-order kinetic equation for this terminal elimination is

$$Ct = C_0' e^{-k_e t,} \tag{9.14}$$

where Ct is the concentration at time t, C_0' is a preexponential term (fictitious concentration at $t = 0$), and k_e is the elimination rate constant.

Linearity of the plot $\ln Ct$ versus time indicates that the model for residue depletion is applicable and linear regression analysis of the logarithmic transformed data can be considered for the calculation of withdrawal periods.

9.5.3.2   Database

The study design must take into account that a minimum of three animals are required at each of a minimum of three slaughter times in the ln-linear phase of the terminal elimination of residues. Anyway, depending on the animal species, four to ten animals per time point are recommended. The FDA (1994) suggests providing residue data of 20 animals with five animals being slaughtered at each of four evenly distributed time points.

9.5.3.3   Theoretical Assumptions

The following assumptions for the linear regression have to be validated:

1. Independence of observations (usually residue depletion meet this assumption because they originate from individual animals)

2. Linearity of the ln-transformed data versus time (the possible lack of fit can be found through an analysis of variance)
3. Homogeneity of variances of ln-transformed data on each slaughter day (Cochran's test is considered the best choice)
4. Normal distribution of the errors (Shapiro–Wilk test for normality should be applied for analysis of residuals)

9.5.3.4 Estimation of Withdrawal Periods by Regression Analysis

Withdrawal periods are defined as the time when the upper one-sided tolerance limit with a given confidence is below the MRL. If this time point does not make up a full day, the withdrawal period is to be rounded up to the next day.

The calculation of the one-sided upper tolerance limit (95 or 99%) with 95% confidence according to Stang (1971) is suggested by the CVMP. Nevertheless, for statistical reasons, the selection of 95% of the tolerance limit (with a 95% confidence level) for setting withdrawal periods should be preferred (see the EMEA/CVMP/036/95/FINAL document for details on this subject; Anon, 1996).

9.5.3.5 Case Study

The data used in this section were taken directly from the EMEA/CVMP/036/95/FINAL document (Anon, 1996). To exemplify the situation, only a subset of the original data (fat tissue) was analyzed.

Data were constructed from an empirical residue depletion study on cattle treated subcutaneously with a veterinary drug. The marker residue in the fat is listed in Table 9.4 and the corresponding MRLs have been set at 20 μg kg^{-1}. The data below the limit of detection were set to half of the detection limit.

To simplify the calculation of withdrawal periods for meat, the CVMP strongly recommend the use of a user-friendly tool (WT1.4 software program) which is merely a computer translation of the method described in the EMEA/CVMP/036/95/FINAL document (Anon, 1996), and is not a new method of its own (downloadable from http://www.emea.europa.eu/exeprogs/vet/56302.zip).

The input data used with the WT1.4 can be seen in Fig. 9.6 and the results obtained are shown in Fig. 9.7, where it is clear that the linearity (F test) and the homogeneity of variances (Cochran test) fulfill the linear regression assumptions. Nevertheless, the normality of the errors was not confirmed (Shapiro–Wilk test). The withdrawal period was estimated as 30 days for the full dataset (95% tolerance limit with 95% confidence interval). Figure 9.8 indicates animal 13 as the possible reason for the nonnormality of the distribution of the residuals obtained. So, animal 13 was excluded because the normality assumption was violated (classified as a possible outlier).

Again, the corresponding subset of data used with WT1.4 can be seen in Fig. 9.9 and the results obtained are shown in Figs. 9.10 and 9.11, where it is clear that the

Table 9.4 Individual results for the marker residue in cattle

Animal number	Days after dose	Fat concentrations ($\mu g\ kg^{-1}$)
1	7	96.8
2	7	225.0
3	7	213.8
4	7	48.3
5	7	119.3
6	7	204.8
7	7	157.5
8	7	450.0
9	7	65.3
10	7	195.8
11	7	148.5
12	7	202.5
14	14	< 2.0
13	14	11.3
15	14	78.8
16	14	51.8
17	14	33.8
18	14	24.8
19	14	2.3
20	14	15.8
21	14	51.8
22	14	13.5
23	14	22.5
24	14	42.8
25	21	27.0
26	21	9.0
27	21	6.8
28	21	6.8
29	21	6.8
30	21	11.3
31	21	40.5
32	21	9.0
33	21	4.5
34	21	9.0
35	21	9.0
36	21	< 2.0
37	28	4.5
38	28	4.5
39	28	9.0
40	28	6.8
41	28	<2.0
42	28	4.5
43	28	<2.0
44	28	<2.0
45	28	4.5
46	28	9.0
47	28	13.5
48	28	< 2.0

linearity (*F* test), the homogeneity of variances (Cochran test), and the normality of the errors (Shapiro–Wilk test) fulfill the linear regression assumptions. Finally, the withdrawal period was estimated as 29 days (95% tolerance limit with 95% confidence interval), which is the final value if the guideline is correctly applied (Anon, 1996).

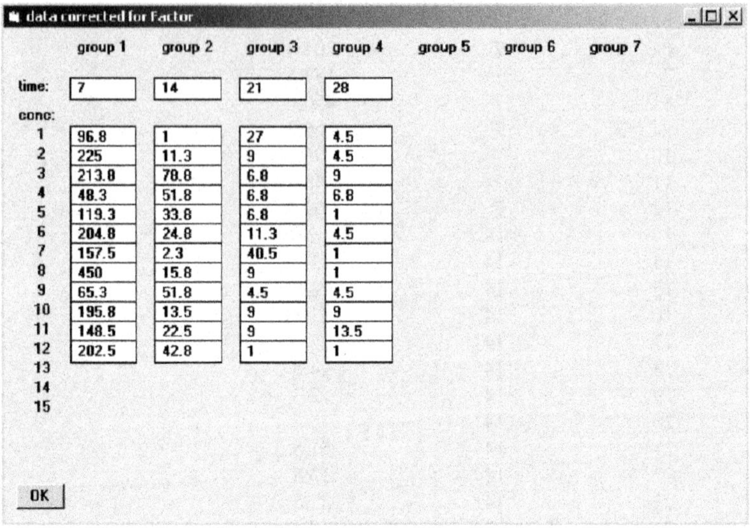

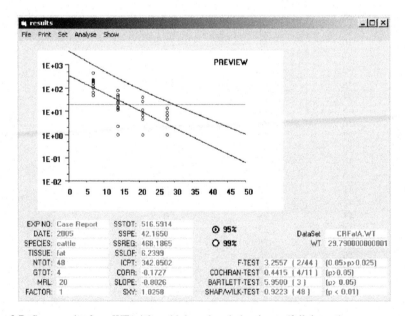

Fig. 9.6 Screen print from the WT1.4 software program for a full dataset

Fig. 9.7 Screen print from WT1.4 for withdrawal period estimate (full dataset)

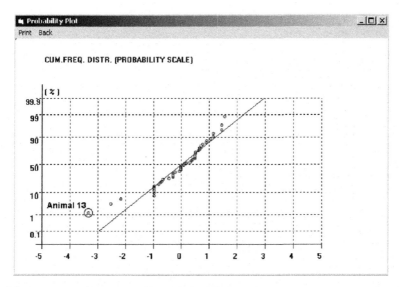

Fig. 9.8 Screen print from WT1.4 for residuals (full dataset)

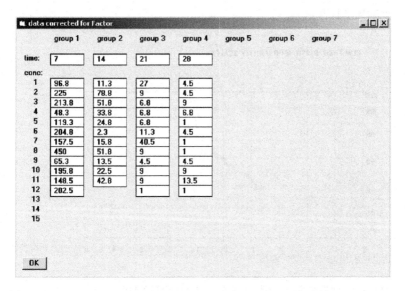

Fig. 9.9 Screen print from WT1.4 for a subset of data (excluding animal 13)

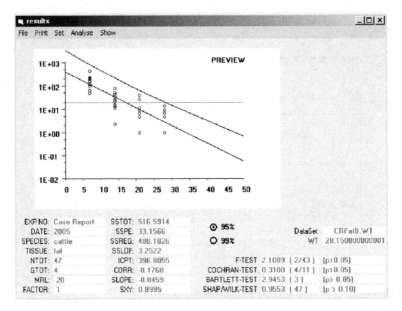

Fig. 9.10 Screen print from WT1.4 for withdrawal period estimate (excluding animal 13)

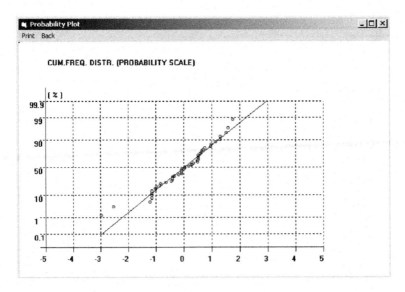

Fig. 9.11 Screen print from WT1.4 for residuals (excluding animal 13)

9.5.4   Withdrawal Period for Milk

The presence of residues in milk can be problematic for human health in multiple ways and the prevention of antibiotic residues is an important aspect of milk quality. Originally, the primary reason for eliminating drug residues from milk was to prevent severe reactions in humans who might be allergic to the drug (β-lactams were the drugs that were of primary concern). Although these reactions do not commonly occur, their severe consequences warrant significant attention. Other factors that make antibiotic residues undesirable in milk include the creation of an off-flavor in milk or subsequent products and, in addition, residues may also interfere with the normal production process for some cheeses by inhibiting or killing bacteria necessary for cheese production (Chester et al., 1996).

The withdrawal period determination for milk shares some assumptions and concepts with the corresponding procedure for meat. However, because the character of milk depletion data and the statistical aspects of calculations with these data differ from those of meat residue data, a separate methodological approach is necessary and an individualized *Note for Guidance for the Determination of Withdrawal Periods for Milk* for the determination of withdrawal periods for milk is available (Anon, 2000).

In fact, before anything else, the withdrawal period definition for milk needs to be contextualized. For example, a milk withdrawal period of 108 h means that all milk up to and including the last milking before 108 h after treatment must be discarded. Depending on the time of treatment in the 12-h milking cycle, the last milk to be discarded may be from the milking at any time point at or after 96 h after the treatment but earlier than 108 h after the treatment. In this example, milk from the first milking at or after 108 h is considered safe. Similarly, a milk withdrawal period of 12 h means that all milking within a 12-h period from the last treatment must be discarded and only milk taken at or after 12 h is considered safe (Anon, 2000).

Regarding methodology, the proposal of the CVMP for the determination of withdrawal periods for milk is the time to safe concentration (TTSC) method. With the TTSC method (fully described in the EMEA/CVMP/473/98/FINAL document; Anon, 2000), tolerance limits on the number of milkings per animal, which is necessary for the residue concentration in the milk of most animals to reach the safe concentration (i.e., the MRL), are calculated. The method assumes a normal distribution after transformation of measuring values onto the natural logarithmic scale (of individual times to safe concentration). In order to derive the TTSC points for each individual animal, monotonic regression preprocessing of the dataset is applied. In a second monotonic regression step the relation between MRL and resulting withdrawal period is smoothed. In accordance with the position already taken with respect to the calculation of withdrawal periods for meat (Anon, 1996), it is recommended to calculate the withdrawal period as the 95/95 tolerance limit, i.e., the upper 95% confidence limit of the 95th percentile of the population.

To avoid the complexity of the withdrawal period calculation for milk (which could be a source of error), the CVMP support the use of appropriate software

(WTM1.4) specifically developed in accordance with the EMEA/CVMP/473/98/ FINAL document (downloadable from http://www.emea.europa.eu/exeprogs/ vet/23100rev1.zip) (Anon, 2000).

9.6   Conclusion

The use of antibiotics in veterinary medicine is essential to prevent or treat diseases. In addition, the role of antibiotics for production purposes is acceptable and desirable whenever they are administered in accordance with guidelines and under qualified supervision. Clinical pharmacokinetics concepts allow us to understand the behavior of drugs when administered in animals. Therefore, MRLs should be fixed on the basis of relevant toxicological data, including information on absorption, distribution, metabolism, and excretion. Nowadays the determination of withdrawal periods seems to be the best option to prevent the transmission of antibiotics (and their metabolites) from animal to humans. Several guidelines that support the calculation of withdrawal periods based on MRLs were developed and released by the authorities.

References

Anadón A (1990). Les residues de substances chimiques dans les aliments d'origine animale en Espagne. B. Acad. *Vet. France.* 63: 245–252

Anadón A; Martínez-Larrañaga MR (1992). Pharmacology and toxicology of quinolones. In: Sunshine, I. (ed). Recent Developments in Therapeutic Drug Monitoring and Clinical Toxicology, Marcel Dekker, New York, pp. 193–198

Anadón A; Martínez-Larrañaga MR (1999). Residues of antimicrobial drugs and feed additives in animal products: *regulatory aspects. Livestock Production Science* 59: 183–189

Anon (1996). Committee for Veterinary Medicinal Products. Note for Guidance: Approach Towards Harmonisation of Withdrawal Periods. (EMEA/CVMP/036/95/FINAL)

Anon (2000). Committee for Veterinary Medicinal Products (2000). Note for Guidance for the Determination of Withdrawal Periods for Milk. (EMEA/CVMP/473/98/FINAL)

Anon (2001). Committee for Veterinary Medicinal Products (2001). Note for Guidance on the Risk Analysis Approach for Residues of Veterinary Medicinal Products in Food of Animal Origin. (EMEA/CVMP/187/00/FINAL)

Anon (2003). Committee for Veterinary Medicinal Products (2003). Guideline on Pre-Authorisation Studies to Assess the Potential for Resistance Resulting from the Use of Antimicrobial Veterinary Medicinal Products. (EMEA/CVMP/244/01/FINAL)

Anon (2004). National Office of Animal Health. (2004). Antibiotics for Animals: An Overview. Briefing Document No. 6

Birkett DJ (1998). Pharmacokinetics Made Easy, McGraw-Hill, Australia

Chester T; Deluyker H; Aerts R; Koopman P, Davot JL (1996). Further considerations in setting milk withholding times for antimicrobials. In: Haagsma N; Ruiter A (eds). Residues of Veterinary Drugs in Food. Proceedings of the Eu-roresidue III Conference, Veldhoven, 6–8 May, pp. 326–330

European Commission (2003). Notice to Applicants and Note for Guidance: Establishment of Maximum Residue Limits (MRLs) for Residues of Veterinary Medicinal Products in Foodstuffs of Animal Origin

FDA (1994). General Principles for Evaluating the Safety of Compounds Used in Food-Producing Animals

Gabrielsson J; Weiner D (2000). Pharmacokinetic and Pharmacodynamic Data Analysis: *Concepts & Applications*, 3rd edn. Swedish Pharmaceutical Press, Stockholm, Sweden

Jones PGH (1997). Regulatory harmonization for veterinary medicines in the European Union. In: Anadón, A.; McKellar, Q. (eds.). Proceedings of the 7th EAVPT International Congress, Madrid (Spain), 6–7 July. J. Vet. Pharmacol. Ther. 20(Suppl. 1): 1–9

Roland M; Tozer TN (1995). Clinical Pharmacokinetics: Concepts and Applications, 3rd edn. Lippincott Williams & Wilkins, USA

Shargel L; Yu ABC (1999). Applied Biopharmaceutics & Pharmacokinetics, 4th edn. McGraw-Hill, NY, USA

Stang K (1971). Angewandte Statistik, Springer, Berlin. Vol. II, pp. 141–143

Vanic ML; Marangunich L; Fernández Courel H; Fernández Suárez A (2003). Estimation the withdrawal period for veterinary drugs used in food producing animals. *Anal. Chim. Acta*. 483: 251–257

WHO. (1989). Gudelines for Predicting Dietary Intake of Pesticide Residues. Geneva

WHO. (1997). The Medical Impact of the Use of Antimicrobial in for Animals. Berlin (Germany), 13–17 October. (Report of WHO Meeting)

Chapter 10
Transport and Reactions of Pollutants

Vassilis Gekas and Ioanna Paraskaki

10.1 Introduction

The aim of this chapter is to provide the food scientist and engineer with tools for understanding the principles of transport and reaction of pollutants and their fate after being released or deposited into the environment. Furthermore, on the grounds of this understanding of basic principles, the food scientist and engineer will possess the ability to model these processes. Mathematical modeling nowadays is facilitated through the use of appropriate computer software programs. There are, generally speaking, a large number of programs available for such modeling and especially for the prediction of the fate of pollutants. When working with these programs it is advisable to understand the principles behind the program rather than treating it as a black box.

This chapter will focus on one of the aspects (phases) of the environment or soil and aquifers such as underground water deposits. Although some comments will be given with a global application, the scope of the present chapter is clearly defined and limited to the abovementioned environmental aspects.

10.1.1 *Atmophilic Versus Lithophilic Pollutants*

Pollutants are chemical substances, elements, or compounds that should be well defined and are a part of the environment that can be chemically defined, and have a specific affinity for certain phases, or parts of the environment which are physically defined. A component which after its release somewhere in the environment preferably passes into the atmosphere is called "atmophilic," otherwise it is characterized as "lithophilic" or "hydrophilic." The quantification of this initial process is

V. Gekas (✉)
Department of Environmental Engineering, Technical University of Crete,
Polytechnioupolis, 73100 Chania, Crete Greece

R. Costa, K. Kristbergsson (eds.), *Predictive Modeling and Risk Assessment,*
DOI: 10.1007/978-1-387-68776-6, © Springer Science+Business Media, LLC 2009

defined by thermodynamics. If we look at the environmental phases, for the purposes of this chapter, we will initially divide the environment into atmosphere and lithosphere (lithosphere in an extended meaning, nonatmospheric) and then we will further divide the lithosphere into a solid phase (soil or the lithosphere in a strict meaning) and the hydrosphere (surface waters, groundwater). The surface waters are the seas, lakes, and rivers. Groundwater is closely related to terms that are frequently used in this chapter or the aquifer and the aquiclude. Let us use the antiknocking agent tetraethyl lead (TEL), a pollutant from automobile engines, as an example. The compound is emitted by automobiles into the environment both in small and in large cities or communities. The question for a food scientist is whether this chemical released in large quantities affects the quality of foods consumed somewhere in these communities in the long run.

In the following sections we will systematically ask questions to try to clarify this fundamental question.

10.1.2 A Prelude to Modeling of Transport in Groundwater

10.1.2.1 The Partition Coefficient: The UNIFAC Model

The first question to ask is whether TEL is an atmophile or a lithophile. Or better to what extent is TEL atmophilic so that it will pass into the atmosphere rather than remain in the lithosphere. Like every component, TEL will be distributed to the two phases. The modeling of this distribution is given by the partition coefficient, K, where

$$K_i = \frac{x_{iI}}{x_{iII}}.$$
(10.1)

I and II signify the two phases, i is the component, and x is the molar fraction of i. The calculation of these parameters is possible with the aid of chemical thermodynamics from where it is known that K equals the reciprocal ratio of the activity coefficients of component i in the two phases I and II:

$$K_i = \frac{\gamma_{iI}}{\gamma_{iII}}.$$
(10.2)

There are a wide spectrum of models in chemical thermodynamics available for the calculation of the activity coefficients. Usually, the procedures are implicit because the activity coefficients are functions of their respective molar fractions. In general, those models are empirical or semiempirical. For most of the organic compounds, the UNIFAC model based on molecule structure and intermolecular interaction parameters can be used. A trial-and-error procedure is generally required. The reader could consult a typical chemical thermodynamics textbook such as Smith et al. (1996) or the excellent book about gas and liquids properties by Reid et al. (1988).

10.1.2.2   Transport and Reaction in the Surface Waters (Rivers and Lakes): The Streeter–Phelps Equations

Traces of TEL have been found in the ocean. This means that TEL is partly hydrophilic or lithophilic. On the other hand, the distances are too far to suppose only hydrospheric transport of this chemical. So transport by the winds is also speculated. The value of K, which is a function of pressure and temperature, of course will show us which part will arrive in the ocean through the atmosphere. The rest will be divided into two phases. One part will arrive in the ocean through the surface waters and the other will reach the aquifers through groundwater. It is the latter which it is subject of this chapter. Nevertheless, we shall say some words on modeling surface water transport and reaction.

In this case, the famous Streeter–Phelps equations are usually applied. These equations are simplified forms of the rigorous transport equation but with source terms. One simplification is that instead of writing a system of transport equations for each component, a collective concentration for the sum of the concentrations, the biological oxygen demand (BOD) concentration, is used and denoted as the concentration L. Of course this means that we face the so-called conventional (bio-degradable) pollutants. In this case we consider also a transport–reaction equation of the dissolved oxygen in the surface water receiver. So there is a system of two ordinary differential equations for lakes and rivers, respectively. But nothing hinders us from using the same idea as the Streeter–Phelps equations for the concentration of one pollutant (provided that it is biodegradable) instead for the BOD concentration, L.

For a river, the accumulation term is neglected and we have a steady state for a continuous emission. The concentration then is mainly a function of the distance and only the kinetics degradation term gives variations in time. If the mean velocity of the river is w, and x the prevailing dimension, then

$$w\frac{\mathrm{d}L}{\mathrm{d}x} = -k_{\mathrm{d}}L, \tag{10.3}$$

where L is the BOD concentration (dimensions ML^{-3}), w is the mean velocity of the river (dimensions LT^{-1}), x is the prevailing coordinate direction of the river (dimension L), and k_{d} is the kinetic constant of the degradation (dimension T^{-1}).

An equation is also needed for the dissolved oxygen, for which the oxygen deficit concentration, D, is conveniently used ($D = DO_{\text{saturated}} - DO$; DO, the dissolved oxygen concentration, is reduced from its initial saturation value in the surface water; dimensions ML^{-3}):

$$w\frac{\mathrm{d}D}{\mathrm{d}x} = k_{\mathrm{d}}L - k_{\mathrm{a}}D, \tag{10.4}$$

where k_{a} is the constant of reaeration.

A similar form of the two equations is used when water forms a lake but then time will be the independent variable instead of the space coordinate x:

$$\frac{dL}{dt} = -k_d L \tag{10.5}$$

and

$$\frac{dL}{dt} = k_d L - k_a D. \tag{10.6}$$

If substance i is nonbiodegradable, for example, a toxic one, then its concentration Ci is used and all source terms implying oxygen are zero in this case.

10.1.3   A Short Introduction to Basic Geology

The basic concept here is the aquifer, which is a saturated permeable geologic unit that can transmit significant quantities of water. An aquifer is confined when it lies between two impermeable layers defined as the aquicludes or aquitards, which are relatively low permeability layers.

An aquifer is anisotropic if it presents different hydraulic conductivities towards the bulky aqueous flow or different coefficients of dispersion or dispersivities towards a given component in different directions owing to a preferential flow direction of the influent in the porous media. The type of aquifer or aquitard material is of importance for the modeling as will be come evident below. Materials such as gravel, sand, silt, and clays are more porous. Limestone is less porous and crystalline rocks are almost impermeable (Table 10.1).

The water table is the surface where the fluid pressure is equal to the atmospheric pressure and a superficial aquifer is one where the water table forms the upper boundary of the aquifer.

Table 10.1 Typical hydrogeologic parameters for various aquifer materials

Materials	Porosity (%)	Specific yield (%)	Hydraulic conductivity (K, cm s^{-1})	Permeability (k, cm^2)
Unconsolidated deposits				
Gravel	25–35	25–35	1–100	10^{-5}–10^{-3}
Sand	30–45	25–40	10^{-4}–10^{-1}	10^{-9}–10^{-6}
Silt	35–45	20–35	10^{-6}–10^{-4}	10^{-11}–10^{-9}
Clay	40–55	2–10	10^{-9}–10^{-6}	10^{-14}–10^{-11}
Rocks				
Karst limestone	15–40	10–35	10^{-4}–10^{-1}	10^{-8}–10^{-6}
Nonkarst limestone	5–15	2–10	10^{-6}–10^{-4}	10^{-11}–10^{-9}
Sandstone	10–25	5–15	10^{-7}–10^{-4}	10^{-12}–10^{-9}
Shale	0–10	0–5	10^{-11}–10^{-7}	10^{-16}–10^{-13}
Crystalline rock (fractured)	1–10	1–10	10^{-6}–10^{-4}	10^{-11}–10^{-9}
Crystalline rock (unfractured)	0–2	0–1	10^{-11}–10^{-9}	10^{-16}–10^{-14}

Water flows very slowly underground, with a rate on the order of magnitude of approximately 1 cm day^{-1}, and it takes a long time for the contaminants to reach a drinking water aquifer. But, on the other hand, once an aquifer has been contaminated it will take a long time to restore it. Sources of groundwater contamination are, for example:

- Agrochemical infiltration (fertilizers, pesticides)
- Infiltration from pits, ponds, lagoons, chemicals stored above ground
- Landfill leachates
- Leaks or spills (fuels, industrial solvents, etc.)
- Nuclear waste
- Pollutants facilitated by precipitation to reach groundwater

10.1.4   Modeling Requirements for the Groundwater Transport and Reaction

In the next sections we will start with modeling the movement of water using the Darcy model. Next, a simplified form of the Navier–Stokes equations should be considered for the bulky flow. Then, to complete the modeling the mass transfer equations for the contaminant(s) with or without reaction or sorption terms will be used.

10.2   Darcy's Law

Understanding the basics of fluid dynamics implies the necessity to start from this basic law which is a constitutive (flux/driving force relationship) equation for the fluid volume transport and it is the heart of the Navier–Stokes equations, which when solved could give the velocities as functions of time and space in a given control volume such as the aquifer is.

From fluid mechanics and transport phenomena in general we know that the driving force in fluid transport is a pressure gradient or, in units of length, the hydraulic head gradient. Indeed if we divide a given pressure by g (acceleration of gravity) we get the pressure in h, the head in length units. This head is generally provided by gravity and pumping but in the case of underground water transport it is the gravity force which gives the pressure which acts as the driving force.

In Fig. 10.1 there is an experimental set-up showing the principle of the Darcy equation. Water flows from a reservoir through a bed from a permeable material. The driving force is $(h_1-h_2)/(x_1-x_2)$ and the flux of volume, which is the apparent velocity of the water flowing through the bed, also called the specific discharge, is given by

$$v_x = -K_x \frac{dh}{dx},$$
(10.7)

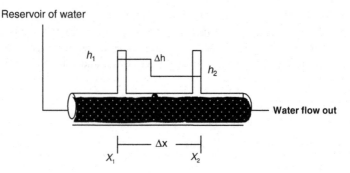

Fig. 10.1 Experimental set-up for the illustration of Darcy's law. (Based on data given by Gekas and Proimaki (1999)

where vx is specific discharge in the x direction (dimensions LT^{-1}), x is hydraulic conductivity in the x direction (dimensions LT^{-1}), and dh/dx is the hydraulic gradient in the x direction (dimensionless).

This depends both on the fluid properties and the aquifer properties. It is greatest in the case of coarse porous materials such as gravel and sands. Values of conductivities for typical aquifer media are given in Table 10.1.

Specific yield is defined as the volume of water that will drain freely per unit decline in the water table per unit surface area of an aquifer. It is dimensionless. The permeability, k, appears in the Darcy equation when pressure is used in the expression for the driving force (dP/dx) divided by the fluid viscosity, μ. Permeability is just an alternative to the hydraulic conductivity to express the capacity of the aquifer material to transmit the water through its body (Table 10.1).

Then the relationship between the two parameters K and k is

$$K = \frac{k\rho g}{\mu}. \tag{10.8}$$

Archie (Gekas and Proimaki, 1999) has used an analogy from the field of electricity between the electrical conductivity and hydraulic conductivity to express this property as a function of porosity, ε^n:

$$K = A\varepsilon^n. \tag{10.9}$$

This is a power law, with A being a proportionality constant.

A popular expression for K in terms of the porosity is the Carman–Kozeny equation (Gekas and Proimaki, 1999), also involving the median particle diameter d_m and the fluid viscosity:

$$K = \frac{\varepsilon^3}{(1-\varepsilon)^2} \frac{d_m^2}{180} \frac{\rho g}{\mu}. \tag{10.10}$$

If the aquifer medium is anisotropic the three-dimensional form of Darcy's law should be used:

$$v_x = -K_x \frac{dh}{dx}, \qquad (10.11)$$

$$v_y = -K_y \frac{dh}{dy}, \qquad (10.12)$$

and

$$v_z = -K_z \frac{dh}{dz}. \qquad (10.13)$$

for longitudinal, transverse, and vertical transport, respectively. In an anisotropic medium such as an aquifer the order will usually be $Kx = Ky > Kz$ because flow in the horizontal plane is preferred over vertical flow through bedded deposits.

10.3   Navier–Stokes Equations

Starting from Darcy's equations in the three dimensions and considering the conservation law, i.e., the continuity equation, and supposing incompressibility (fluid density constant), we get a simplified partial differential equation in three dimensions:

$$\frac{\partial}{\partial x}\left(K_x \frac{\partial h}{\partial x}\right) + \frac{\partial}{\partial y}\left(K_y \frac{\partial h}{\partial y}\right) + \frac{\partial}{\partial z}\left(K_z \frac{\partial h}{\partial z}\right) = S\frac{\partial h}{\partial t}, \qquad (10.14)$$

where h and the K have been defined previously and S is the specific storage or the volume of water which a unit volume of the aquifer releases from storage under the unit decline in hydraulic head (with dimensions L^{-1}).

S is estimated from two parameters: the aquifer compressibility α and the fluid compressibility β:

$$S = \rho g(\alpha + \varepsilon\beta). \qquad (10.15)$$

The dimensions of the compressibilities are $LT^2 M^{-1}$.

We can further simplify the flow equation by neglecting the vertical transport compared with the horizontal transport and if steady-state conditions are assumed the equation may be reduced to the well-known Laplace equation in two dimensions:

$$\frac{\partial^2 h}{\partial x^2} + \frac{\partial^2 h}{\partial y^2} = 0. \qquad (10.16)$$

10.4   A Numerical Solution of the Laplace Equation

Boundary equations will naturally be required. Let us look at the following data for the aquifer: $K = 100$ L day^{-1} dm^{-2} (isotropic), $H_0 = 500$ m (initial head), $\Delta x = \Delta y = 7$ m, and $dh/dx = 0.05$ m/1 m (hydraulic gradient; slope). We wish to calculate the heads for a simple two-dimensional aquifer (depth and longitude) assuming no-flow boundary at the bottom. We use a five point method, central differentiating technique. In this case we use 13 nodes in the longitudinal i-direction and seven nodes in the vertical j-direction. Then the head in an ij node will be

$$h_{ij} = \frac{h_{i,j+1} + h_{i,j-1} + h_{i-1,j} + h_{i+1,j}}{4} \qquad (10.17)$$

for an isotropic aquifer and if the hydraulic permeability KX (Kx) is different from KY (Ky), then the discretization gives

$$h_{ij} = \frac{KX_{i,j}h_{i+1,j} + KX_{i-1,j}h_{i-1,j} + KY_{i,j}h_{i,j+1} + KY_{i,j-1}h_{i,j-1}}{KX_{i-1,j} + KX_{i,j} + KY_{i,j-1} + KY_{i,j}}. \qquad (10.18)$$

A Fortran program, FLOWNET, has been written to solve the system of equations arising and to calculate the flow rate Q with variable hydraulic conductivity (Schnoor, 1996).

Program Flownet

```
C       EXAMPLE TO CALCULATE FLOW 'Q' WITH VARIABLE PERMEABILITY
        REAL*4 H(13, 7), KX(13,17),KY(13,7), Q(13), H0, DX, SLOPE, KK,KO
1       WRITE (6.5)
5       FORMAT <8 'ENTER INITIAL HEAD; VALUE FOR DX; AND SLOPE:'
        READ (5,10) HO; DX; SLOPE
        FORMAT (F10.2/F10.2/F10.2)
C       DEFINE THE KX AND KY ARRAYS
        KO=100
        DO 15 J=1,7
        DO 15 I=1,13
        KX (I,J)= KO
        KY (I,J)=KO
15      CONTINUE
C       INITIALIZE ALL HEADS TO HO
        DO 20 J=1,7
        DO 20 I=1,13
        H (I,J)=HO
20      CONTINUE
C       WATER TABLE BOUNDARY: H(X, Y) =HO-(HO-H)*(X/XL)
        DO 30 I=2, 12
```

```
            H (I,1) = SLOPE*DX*(I-2)+HO
30          CONTINUE
C           KEEP TRACK OF THE NUMBER OF ITERATIONS AND OF LARGEST ERROR
C           NO FLOW BOUNDARIES NEED TO BE RESET WITHIN EACH ITERATION
               LOOP
            NUMIT=0
40          AMAX =0.0
            NUMIT =NUMIT+1
C           LEFT AND RIGHT NO FLOW BOUNDARIES
            DO 50 J=1,7
            H (1, J)=H(3, J)
            H (13, J)=H(11, J)
50          CONTINUE
C           BOTTOM NO FLOW BOUNDARY
            DO 60 I=2,12
            H(I, 7)= H(I,5)
60          CONTINUE
C           SWEEP INTERIOR POINTS WITH 5-POINT OPERATOR
            DO 70 J=2, 6
            DO 70 I=2, 12
            OLDVAL=H(I, J)
            KK= KX(I-1, J)+KX(I, J)+ KY(I, J-1)+KY(I,J)
            H(I, J)= KX(I-1, J)*H(I-1, J)+KX(I, J)*H(I+1,J)+KY(I, J-1)*H(I, J-1)+
               KY(I, J)*H(I,J+1)
            H(I, J)= H(I, J)/KK
            ERR=ABS(H(I,J)-OLDVAL)
            IF (ERR.GT.AMAX) AMAX=ERR
70          CONTINUE
C           DO ANOTHER ITERATION IF LARGEST ERROR AFFECTS FIRST
               DECIMAL POINT
            IF (AMAX.GT.0.001) GO TO 40
C           WE ARE DONE
            DO 75 I=2,12
75          CONTINUE
            WRITE (6,80) NUMIT, ((H(I, J), I=2,12), J=1,6)
80          FORMAT (///1X, 'NUMBER OF ITERATIONS IS', I5,/// 6(11F8.1///))
            WRITE (6,90) (Q(I), I=2,12)
90          FORMAT ('FLOW RATE ACROSS THE WATER TABLE IS:", /(11F8.2)
```

```
        WRITE (6,100)
100     FORMAT (// DO YOU WISH ENTER MORE DATA? IF SO ENTER "YES",')
        READ (5, 105) ANS
105     FORMAT (A2)
        IF (ANS.EQ.'YE') GO TO 1
        CALL EXIT
        END
```

Running the program the following appears:

ENTER INITIAL HEAD; VALUE FOR DX AND SLOPE:

Then we give the values:

500.00

7.00

0.05

10.5   The Pollutant Solute Transport Equation

Fick's second law and the mass-balance components in a given control volume of a homogeneous medium will give the solute transport equation, as is well known from the transport phenomena theory.

The equation consists of an accumulation term, a convection term, dispersion (instead of diffusion) terms, and the source (physical, chemical, biological reaction) terms. In one dimension with x as the prevailing direction we have

$$\frac{\partial C}{\partial t} + u_x \frac{\partial C}{\partial x} = E_x \frac{\partial^2 C}{\partial x^2} \pm \sum_{m=1}^{n} r_m. \tag{10.19}$$

In its three-dimensional form the equation is as follows in gradient operator notation

$$\frac{\partial C}{\partial t} + u_i \nabla C = E_{ij} \nabla^2 C + \sum_{m=1}^{n} r_m, \tag{10.20}$$

where C is the pollutant concentration (dimensions ML^{-3}), t is time (dimension T), and u_i is the the velocity vector (u_x, u_y, u_z). Those velocities are the volume fluxes and are estimated through the simplified Navier–Stokes equations (Sects. 10.2, 10.3) and have dimensions of LT^{-1}. E_{ij} is the dispersion coefficient tensor:

$$E_{ij} = \begin{vmatrix} E_{xx,} E_{xy} E_{xz} \\ E_{yx} E_{yy} E_{yz} \\ E_{zx} E_{zy} E_{zz} \end{vmatrix}.$$

At this point we have to remind the reader that the hydrodynamic dispersion is orders of magnitude more important in the field than molecular diffusion. Mathematically this is like molecular diffusion (Gekas, 1992).

Because the medium is anisotropic we cannot exclude the possibility that a driving force in one direction, for example, in the x direction, does not cause a flux in another direction, for example, in the y direction. In this constitutive equation the cross-coefficient E_{xy} is used.

In the case the cross-coefficients are too small compared with the conjugative coefficients E_{xx}, E_{yy}, and E_{zz} and sometimes in the literature the E-tensor reduces to an E-vector E_i (E_x, E_y, E_z). But we should not forget the correct principles and we are always aware of the presuppositions made.

Values for the dispersion coefficients can be obtained from the literature or can be obtained by tracer injection experiments in the aquifers. Dispersion does not have a turbulent origin. Groundwater flow is laminar. The groundwater flow through the porous media follows rather tortuous pathways. Microscopic and macroscopic variations lead ultimately to dispersion.

There is then a dependence of E on the prevailing velocity u. The coefficients are called dispersivities and have units of length. This is valid both for the longitudinal and the lateral dispersion:

$$E_x = \alpha_x u_x + D_0 \tag{10.21}$$

and

$$E_y = \alpha_y u_x + D_0. \tag{10.22}$$

The order of magnitude of the longitudinal α depends on the scale where the phenomena take place. Laboratory values are smaller, on the order of 0.01 m, and field scales are on the order of 25 m, the range being from 1 to 100 m. For lateral dispersion the values are smaller usually than for longitudinal dispersion. Knowing the dispersivities, one can calculate E values and solve the pollutant equation. But care should be taken also with regard to the source terms.

10.6   The Source Terms

10.6.1   Physical: Sorption, Hydrophobic Retardation

We characterize physical terms as those arising from physical sorption and physico-chemical or thermodynamic phenomena, but as chemical terms we only consider chemical reactions in which the pollutant molecule participates.

First we expect that the pollutant molecule will be forced to leave the aqueous phase by a number of different mechanisms to be attached onto the particles of the aquifer medium. Generalizing to include all physicochemical mechanisms, we define a distribution coefficient K_d. Then we accept a linear relationship for the typical low pollutant concentrations C in the groundwater:

$$S = K_d C,$$ (10.23)

where S is the amount sorbed (mg of pollutant/kg of porous medium) as a function of the concentration C in the aqueous solution (mg L^{-1}). Then the distribution coefficient has units of liters per kilogram. Values of K_d are given in Table 10.2.

Values are also given for the partition coefficient K_{ow} for various pollutants for the pair water–octanol to express the hydrophobicity of organic pollutants (Table 10.3). Then the source term for sorption becomes

$$r_{sorption} = -\frac{\partial S}{\partial t} \times \frac{\rho_b}{\varepsilon},$$ (10.24)

where the partial time derivative of S is multiplied by the bulk density and divided by the porosity of the aquifer material.

The meaning of the symbols K_{ow} and K_{oc} in Table 10.3 are the following: K_{ow} is the water–octanol partition coefficient and K_{oc} is the organic carbon content normalized partition coefficient. There is an empirical relationship between K_{ow} and K_{oc}. In general

$$\log K_{oc} = a \log K_{ow} + b.$$ (10.25)

Examples of parameters a and b are given in Table 10.4 for pesticides.

10.6.2   Chemical and Biochemical Reactions

We consider the redox reactions both for organics and for metals as crucially important. A typical organic containing C, H, and O, such as CH_2O, will be oxidized according to the redox potential of the groundwater. The primary electron donor is

Table 10.2 Orders of magnitude values of K_d in sandy aquifers for toxic metals

Toxic metal	K_d (L kg^{-1})
As	1–10
Cd	5–50
Se	1–10
Cr	10–100
Cu	5–50
Pb	5–50
Zn	2–25
^{239}Pu	50–500
^{60}Co	10–100
^{90}Sr	5–50

Table 10.3 Parameters K_{ow} and K_{oc} and solubility according to Schnoor (1996)

Component	log K_{ow}	log K_{oc}	Solubility in water (20°C; mg L^{-1})
Bromoform	2.30	2.16	–
CCl$_4$	2.64	2.40	800
Chloroethane	1.49	1.57	5,740
Chloroform	1.97	1.92	8,200
Chloromethane	0.95	1.18	6,450
CCl$_2$F$_2$	2.16	2.05	280
CCl$_3$H$_2$	1.26	1.41	2,000
C$_2$H$_6$	3.34	2.81	50
C$_2$Cl$_4$	2.88	2.57	200
C$_2$HCl$_3$	2.29	2.15	1,100
VC	0.60	0.93	90
1,1-C$_2$H$_4$Cl$_2$	1.80	1.80	400
1,1-C$_2$H$_2$Cl$_2$	1.48	1.57	400
1,2-C$_2$H$_4$Cl$_2$	1.48	1.57	8,000
1,2-trans-C$_2$H$_2$Cl$_2$	2.09	2.00	600
1,1,1-C$_2$H$_3$Cl$_3$	2.51	2.30	4,400
1,1,2-C$_2$H$_3$Cl$_3$	2.07	1.99	4,500
Aromatics			
Benzene	2.13	1.92	1,780
Benzo[a]pyrene	6.06	5.85	0.0038
Chlorobenzene	2.84	2.63	500
Ethylbenzene	3.34	3.13	152
Hexachlorobenzene	6.41	6.20	0.006
Naphhtalene	3.29	3.08	31
Nitrobenzene	1.87	1.66	1,900
Pentachlorophenole	5.04	4.83	14
Phenole	1.48	1.27	93,000
Toluene	2.69	2.48	535
1,2-Dichlorobenzene	3.56	3.35	100
1,4-Dichlorobenzene	3.56	3.35	79
1,2,4-Triclorobenzene	4.28	4.07	30
2-Chlorophenole	2.17	1.96	28,500
2-Nitrophenole	1.75	1.54	2,100
2,4,5,2 ,4 , 5 -PCB	6.72	6.51	–
Pesticides			
Acroleine	0.01	0.86	210,000
Alachlor	2.92	2.96	242
Atrazine	2.69	2.55	33
Dieldrin	3.54	3.33	0.2
DDT	6.91	6.70	0.0055
Lindane	3.72	3.51	7.52
2,4-D	1.78	1.65	900

Table 10.4 Examples of parameters a and b for Eq. 10.25

Pesticide	a	b	R^2
Wide variety	0.544	1.377	0.74
Dinitroaniline herbicides	0.937	0.006	0.95
2-chlorinated	1.00	−0.21	1.00
Insecticides, fungicides	1.029	−0.18	0.91
Halogenated aliphatic and aromatic	1.4	0.5	0.95

the dissolved oxygen in the groundwater. Then other species follow: dissolved oxygen > nitrates > manganates > ferrous iron+> sulfates > carbon dioxide.

The change of free enthalpy for the oxidation reactions has a negative value, indicating spontaneity, although the rate at which these reactions occur may not be significant.

Michaelis–Menten kinetics may be used to model the source term of the biodegradation of an organic pollutant by oxygen:

$$r_{redox} = \frac{X}{Y} \frac{\mu S}{K_s + S} \left(\frac{[O_2]}{K_{O_2} + [O_2]} \right),$$ (10.26)

where r is the rate dS/dt (dimensions $ML^{-3} T^{-1}$), t is the time (dimension T), S is the substrate concentration of organics (dimensions ML^{-3}), X is the viable biomass (dimensions ML^{-3}), Y the biomass yield (mass of cells/mass of substrate), μ is the maximum biomass growth rate (dimension T^{-1}), K_s is the half-saturation constant, $[O_2]$ is the oxygen concentration in terms of partial pressure in atmospheres, and is the half-saturation constant for oxygen as an electron acceptor (0.1 atm).

Equation 10.26 expresses organic biodegradation under low oxygen conditions that occur in the groundwater. Some organics can be degraded only under aerobic and some only under anaerobic conditions. For example, aromatic compounds are readily degraded aerobically although the process can be slowed down in the groundwater, but halogenated aliphatics are difficult to degrade aerobically. However, there are some steps between the aerobic transformation and the anaerobic one:

- Aerobic oxidation – oxygen is the reactant, CO_2 and H_2O are produced
- Anoxic denitrification – nitrates are the reactants, nitrogen and H_2O are produced
- Anoxic sulfate respiration – sulfates are the reactants, HS, CO_2, and H_2O are produced
- Anaerobic methanogenesis – CO_2 is the reactant, CH_4 is produced

Because of the source term either from the sorption or hydrophobic effect or a chemical, the transport and the reaction of the pollutants are retarded and this is explicitly shown in the transport equation by the factor R, the dimensionless retardation factor including K_d, the distribution coefficient, and the porosity and the mass density of the aquifer, ρ_b:

$$R = 1 + \frac{K_d \rho_b}{\varepsilon}.$$ (10.27)

The transport equation becomes

$$\frac{\partial C}{\partial t} + \frac{u_x}{R} \frac{\partial C}{\partial x} = \frac{E_x}{R} \frac{\partial^2 C}{\partial x^2} \pm \frac{1}{R} \sum_{m=1}^{n} r_m.$$ (10.28)

R slows down the entire process of pollutant migration.

10.7   Spills from Landfill and from Nonaqueous Liquid Tanks

10.7.1   Landfill Leachate Spills

Schnoor (1996) gives a good example of a sequence of reactions occurring to a landfill leachate which migrates in a one-dimensional water table aquifer (Table 10.5). The velocity of water is approximately 10 cm day^{-1}. Measurements were made downhill from the aquifer.

The above results can be interpreted considering that, first, dissolved oxygen is consumed for the aerobic degradation of organics. Then iron is reduced from Fe(III) to Fe(II). Subsequently the Fe(II) concentration decreases because of precipitation as FeS or FeS$_2$.

The pH increases because of proton consumption for the reduction of iron and sulfates. In addition to the degradations of organics, heavy metals undergo valence state changes as the redox potential decreases and this affects their tendency for sorption, ion exchange, complexation, and other reactions.

10.7.2   Non-aqueous-phase Liquids

Liquids which are immiscible with water phases occur when there are accidents from spills either from nonaqueous liquid tanks or from landfills. Non-aqueous-phase liquids (NAPLs) that are less dense than water, such as gasoline, crude oil, fuel oil, or jet oil, form a floating pool of material on the surface of the groundwater. Take a spill of 10,000 gal from gasoline as an example. Before this spill reaches the aquifer's saturated zone it must percolate through the unsaturated zone (no water there) and it may be retained by the soil of this unsaturated zone. Usually contamination of the water in aquifers may be expected only when amounts higher than 10,000 gal are released (Schnoor, 1996).

NAPLs that are denser than water (DNAPLs) are transported by gravity through the unsaturated zone and of course a portion of this quantity can by retained in the unsaturated zone. Once DNAPLs reach the saturated zone, plumes of chemicals

Table 10.5 Properties of water leached from landfills

Distance from landfill (m)	E_H (V)	pH	O$_2$ (mg L^{-1})	Fe(II) (mg L^{-1})
0	+0.52	5.6	4.0	<0.1
100	+0.48	6.0	3.0	<0.1
300	+0.05	6.8	0.4	2.0
500	−0.10	7.3	0	6.0
700	−0.15	7.8	0	2.5
900	−0.20	8.3	0	0.2

dissolved in the water may occur and the rest of the DNAPL will tend to descend to the lower part of the aquifer, forming a pool of organic fluid. So, the first remediation task is for an environmental or environmentally oriented engineer to try to collect the pure phases using submersible pumps.

The dissolution rate of NAPLs can be modeled through the transport equation, taking into account a volumetric fraction of the aqueous phase, ϑ_w, and an appropriate source term considering the mass-transfer coefficient, k, between the NAPL phase and the water, multiplied by the difference of the concentrations C_s (that is, the equilibrium concentration of the pollutant in the water in contact with the NAPL) minus C:

$$\vartheta_w \frac{\partial C}{\partial t} + u_x \frac{\partial C}{\partial x} = \vartheta_w E_x \frac{\partial^2 C}{\partial x^2} + k_0 (C_s - C).\backslash] \tag{10.29}$$

10.8 Numerical Solutions

Various finite-difference and finite-element computer software programs have been developed and can be applied to solve the transport–reaction equations referred to in this chapter. One such program is the Fortran program Flownet that is applied to solve the two-dimensional transport model Ifor the aqueous-phase flow (Eq. 10.4). Similar procedures may be followed to solve the pollutants' transport–reaction equations, where finite differences and finite elements may be used.

Food scientists and engineers are familiar with the FINDIFF program to model heat transfer problems occurring in foods. Because of the analogies with mass transfer, it is possible to modify it for mass transfer and in particular for mass transfer of a pollutant.

Another program used by the authors and collaborators is FEMLAB, i.e., a MATLAB tool for solving numerically through finite elements and differences the partial differential equations. It is recommended that the reader understands the principles of numerical methods so that these software packages and tools do not remain a black box.

In the first place we need to know the exact form of the transport–reaction equation that the software has been designed to solve and if necessary we must modify it for our application. Secondly, it is very important to consider the correct initial and boundary conditions. What are the presuppositions assumed by the software and how are these compatible with the actual problem of the user? One needs to understand the discretization techniques. Test the sensitivity by changing the grid parameters, the space steps Δx, Δy and the time step Δt and observe what changes occur. If you have two algorithms solving the same problem, they must be comparable. A typical mistake is that the user forgets to check the validity of balances. The conservation of mass needs to be maintained and understood.

If the reader would like to become familiar with the numerical techniques but has not attended or it is not likely to attend a course on this subject, we recommend the *Basic Environmental Engineering* by Bungay (1988), where examples may be found, for example, for solving the Streeter–Phelps equations, using BASIC. Also, in *Environmental Modeling Fate and Transport of Pollutants in Water, Air, and Soil* by Schnoor (1996) the reader will find algorithms for solving similar problems.

10.9   Uptake by Plants

After considering transport and reactions of the pollutants and their modeling we are at a critical point regarding uptake of pollutants by plants through their roots. This is, of course, crucial for the contamination of foods. Will contamination of foods occur and to what extent? Suppose we found a concentration C_1 of a given pollutant in an area near the roots of a plant, but plants take up chemicals from the soil near their roots. From plant physiology it is known that every living entity, tissues, cells, organelles, etc., is surrounded by a membrane which controls the transport of components to and from this entity. Thus, concerning the uptake of components from the environment, life is protected, to some degree, against invasion, but only to a certain degree. The question is, to what degree?

To answer this question the properties of the biological membranes need to be known. The author belongs to the school of irreversible thermodynamics as far as modeling of membranes is concerned (Gekas et al., 1993). Three parameters can characterize any membrane: L_p, the hydraulic conductivity; σ, the reflection coefficient as the theoretical, the true retention of a given component which is obtained when the membrane works in absence of limiting phenomena; ω, and the component's mobility . Of those three, σ is the most crucial. σ affects both the retention of the membrane and fluxes of both the solvent (water), J_v, and the solute (the pollutant), J_s:

$$J_v = L_p(\Delta P - \sigma \Delta \Pi) \tag{10.30}$$

and

$$J_s = J_v(1 - \sigma)C_s + \sigma \Delta \Pi. \tag{10.31}$$

It is also understood that because of the active transport and consequently the specialized permeabilities due to the protein carrier function, mass transfer through biological membranes is a very complicated phenomenon. So the coupling to ATP hydrolysis should also be taken into account.

With respect to some components, for example, oxygen, carbon dioxides, and most organics, the membrane shows the so-called basal permeability, that is, the passive transport through the phospholipid layer. Other components such as certain ions, such as K^+, Ca^{2+}, and Na^+, glucose, and other monosaccharides present a very low basal permeability and their transport to or from the interior is controlled and/or facilitated by active transport.

To answer to the question which component is transferred by basal permeability and which by specialized permeability intelligently, we can measure the permeability in the lipid bilayer phase and the overall membrane permeability and see whether the two differ from each other. If they are equal, such as for oxygen, carbon dioxide, and glycerol, then those components are transferred only by basal permeability. If their permeability is greater (often orders of magnitude higher) than what it should be if this were shown by the lipid bilayer, such as for water, K^+, Na^+, and Cl^-, then this is an indication that they are moved with the aid of carriers. The diagram in Fig. 10.2 is based on data given by Taic and Zeiger (2003). The points in series 1 lie on the dichotomy of the X–Y diagram; their permeability is equal to the basal one. From bottom to top three chemicals are transported solely by the phospholipid phase; oxygen, carbon dioxide, and glycerol. Points 2 and 3 deviate from the dichotomy and represent water and three ions: K^+ (10, 6), Na^+ (10, 7.5), and Cl^- (10, 8.5). The units for permeability are centimeters per second and both axes show the negative decadic logarithm of the permeability.

Whereas the irreversible thermodynamics approach is a nice first approximation, there is still much that has to be done and research is indeed currently being conducted to tackle this highly challenging subject area. The authors are currently carrying out projects both experimental in cooperation with Lund University and theoretical work is being conducted to present a satisfactory model for mass transfer at a cellular level.

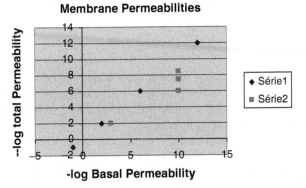

Fig. 10.2 Typical values for the basal and total permeability of a biological membrane towards various components (Based on data from Taic and Zeiger 2003)

References

Bungay, H., 1988, Basic Environmental Engineering, Biline Associates, New York, see for the Streeter Phelps modeling

Gekas, V., 1992, Transport Phenomena of Foods and Biological Materials, CRC Press, see for the Irreversible Thermodynamics model

Gekas, V. and Proimaki, S., 1999, Transport Phenomena for Environmental Engineers, Tziola, Thessaloniki

Gekas, V., Tragårdh, G., and Hallström B., 1993, Ultrafiltration Membrane Performance Fundamentals, The Swedish Foundation for Membrane Technology, Lund University, see for membrane mass transfer and comparison of the various models

Reid, R. C., Prausnitz, J. M., and Poling B. E., 1988, The Properties of Gases and Liquids, 4th edition, McGraw-Hill, Singapore

Schnoor, J. L., 1996, Environmental Modeling Fate and Transport of Pollutants in Water, Air, and Soil, Wiley, New York, see for the strict mathematical modeling and the numerical solutions

Smith, J. M., Van Ness H. C., and Abbott, M. M., 1996, Introduction to Chemical Engineering Thermodynamics, McGraw-Hill, New York

Taiz, L. and Zeiger, E., 2003, Plant Physiology, 3rd Edition, Sinauer Associates, Sunderland, MA, see about cell membrane permeabilities and root uptake of pollutants

Chapter 11
Environmental Pollution from Pesticides

Ioannis T. Polyrakis

11.1 Introduction

The control of crop pests is as old as human civilization itself. Since the prehistoric era, human beings have used several primitive means in an effort to protect their crop yields. The Chaldeans were the first to use jackstraws. In addition, it is widely known that the ancient Greeks and Romans used sulfur, the effectiveness of the use of which, however, seems to have been known even earlier, by the Sumerians around 2500 B.C. The burning of plants was usual in ancient times for the disinfection of closed rooms (Koutselinis, 1997).

In spite of substantial human effort, the evolution of the several means of plant protection has been considerably retarded, and only from the middle of the six-teenth century did man witness the appearance of chemicals for the purpose of controlling certain crop pests. Arsenic was first used in 1669 and tobacco extract (whose active ingredient is nicotine) was first used as an insecticide in 1690. In 1845, the government of Prussia introduced phosphorus as a rodenticide; derris appeared around 1848 as an insecticide and in 1868 Paris green started being applied for the control of potato *Coleopterans*. From then and until the outbreak of World War II, various means of plant protection were used, with inorganic sub-stances prevailing among them. The first organic pesticide was DDT; it had been known since 1874 when it was invented by a German chemist, but only in 1939 did the Swiss Paul Müller ascertain its insecticidal qualities. Since that time many new chemical substances have been produced and nowadays the number of pesticides exceeds 10,000, of which only 600–800 are used, whereas there are about 12,000 commercial combinations. In the Greek market, for example, about 1,350 commer-cial products exist (Polyrakis, 2004).

Pesticides have tremendously benefited mankind by increasing food and fiber production, and by controlling the vectors of serious plant, animal, and human dis-eases and weeds of crops. If pesticides could not be used for pest control, farmers'

Ioannis T. Polyrakis (✉)
Department of Environmental Engineering, Technical University of Crete,
Polytechnioupolis, 73100, Chania, Crete, Greece

R. Costa, K. Kristbergsson (eds.), *Predictive Modeling and Risk Assessment,*
DOI: 10.1007/978-1-387-68776-6, © Springer Science+Business Media, LLC 2009

Table 11.1 Use of pesticides during last years of the 1990s (OECD, 2000)

Country/group of countries	Pesticides (tons/km^2 of agricultural land)
France	0.59
Greece	0.23
Italy	0.78
Korea	1.29
Portugal	0.43
Switzerland	0.37
USA	0.21
European countries of the OECD	0.34
OECD	0.24

production of crops and livestock might decrease by 25–30%. Even so, the use of pesticides has caused considerable negative effects. It has benefited humanity but at the same time has caused unwholesome phenomena and unhealthy conditions. This is because the structure of pesticides and their mode of operation, especially the way they are used by farmers, have rendered them as serious pollutants for many agroecosystems and for the environment in general. Owing to their chemical nature, pesticides are biocides, which have the potential for poisoning organisms other than the target insect, microorganism, or plant species that should be controlled (Lee, 1979).

In Greece, the annual consumption of insecticides is estimated at more than 150,000 tons, excluding sulfur and $CuSO_4$. During the last few years of 1990s, 0.23 tons of pesticides per square kilometer of agricultural land was used (OECD, 2000) (Table 11.1).

Nowadays, the problem of side effects caused by pesticides is a consequence of past ignorance and/or current indifference, in spite of the sensitivity concerning environmental management.

11.2 Pesticides and the Atmosphere

Outside the food chain, the atmosphere is probably the most important medium for long-distance dispersal of pesticides. Drift and evaporation during aerial application, volatilization from crops and agricultural soils, wind erosion of contaminated soils, and emissions from manufacturing and disposal processes constantly add to the level of pesticides in the atmosphere. Prevailing air currents may transport these pesticides and their alteration products over a great distance (Wolfe, 1979; Durham, 1979).

Pesticides belong to the category of semivolatile organic compounds, and can be transported owing to their aerial/particulate distribution into the atmosphere (Valavanidis, 2000). In this manner, pesticides used in one country may reach another in which their use has been banned. Pesticide residues may return from the atmosphere to the surface of Earth by way of rainout, fallout, or direct sorption. In largely agricultural regions, pesticides in the ambient air may constitute a major source of exposure for human beings and wildlife. On a global scale, airborne pesticides may affect the ecological balance in nonagricultural areas.

There follows a description of the atmospheric inputs of pesticides.

11.2.1 Spray Application

Enormous quantities of pesticides are released into the atmosphere during spray application to crops. The quantity of a pesticide that reaches a target depends on a number of factors, including the amount of material sprayed, the physicochemical properties of the pesticide and dispersion vehicle, the particle size distribution, the height at which the material is released, wind speed, and atmospheric turbulence. The particle size distribution of the spray, however, is perhaps the most important factor contributing to off-target drift. The most obvious way to distribute extremely small amounts of pesticide over large surface areas is to reduce the particle size of the droplet so that its total surface area approaches that of the target surface. Assuming a spread factor of 4 times the diameter of an original aqueous droplet and a spray volume of between 15 and 77 L/ha, a drop size of less than 100-mm diameter begins to approach the ideal (Furmidge, 1963). According to Orphanidis (1968), the reduction of the droplets' dimensions must not be smaller than a marginal rate. Actually, droplets with a diameter smaller than 0.5 mm are unusable.

On the other hand, the size of the droplets that finally reach a target surface depends on a number of factors, including nozzle's diameter, the pressure of the liquid at the nozzle, the type of the nozzle, the way of siphoning the liquid, the type of the chemical and of the dispersion vehicle, the distance of the application, the relative humidity and the temperature of the environment (which affect the evaporation), as well as the direction and the velocity of the wind that blows away the smaller droplets. The speed of spray drift is inversely related to droplet size (Table 11.2).

Spray droplets of less than 30-mm volume mean diameter have been shown to drift great distances (Yeo, 1959; James and Vaughan, 1970). Aerosols of 2-mm volume mean diameter have been found to drift as much as 34 km in a 4.8 -km/h wind while falling only 3 m (Akesson and Yates, 1964). Crop damage from aerial herbicide application has been reported as far as 88.5 km downwind from the target area (Akesson, 1973). According to Woodwell et al. (1971) in a case of aerial application of DDT to forests, less than 50% reached the target area.

The drift of small droplets from the target surface is attributed to their swing into the air for long time (Table 11.3).

Furthermore, little is known about the optimum spray drop that should be used for a given pesticide to attain the most effective results. Optimum particle size for achieving effective weed or pest control must be balanced by consideration of aerosol drift, cost, and the environmental impact. The relationship between pesticide droplet size and specific toxicity is inadequately understood (Hurtig, 1973).

Table 11.2 Influence of droplet size on spray drift (Orphanidis, 1968 and his sources)

Diameter of spray droplet (μm)	Spray drift (feet) under constant velocity of the wind (3 miles/h)
5	18,000
33	409
100	48
200	19
500	7
1,000	5
3,000	4
6,000	<4

Table 11.3 The relationship between droplets' size and their wing (Orphanidis, 1968 and his sources)

Droplets' diameter (μ)	Time (s) of wing of droplets ($h = 3$ m)
5	67.5
10	16.9
20	4.2

Vaporization of pesticide compounds during and immediately after spray operations may be a more important mechanism for atmospheric entry than particle drift. Evaporation of liquid aerosol droplets as they fall through the air can release substantial quantities of the pesticides contained therein as vapors or as very small particles (Decker et al., 1950; Kiigemagi and Terriere, 1971).

More than 40% of the original aqueous spray volume from an air-blast sprayer was found by Cunningha et al. (1962) to be lost by evaporation before the droplets had fallen 11 m. On the other hand, Ware et al. (1970) reported that less than 50% of aerially applied pesticides were deposited on agricultural targets in Arizona, during summer months.

11.2.2 Volatilization from Crops and Soils

Volatilization is the physicochemical process by which a compound is transferred to the gas phase. It can result from evaporation from a liquid phase, sublimation from a solid phase, evaporation from an aqueous solution, or desorption from the soil matrix. The transfer of a compound from a compartment (soil or plant) into the atmospheric compartment involves several steps including not only these phase transformations but also transport processes (Bedos et al., 2002).

A major route of entry of pesticides into the atmosphere is through volatilization from treated agricultural soils and plant foliage. This process certainly competes with direct input during spray application. Well over 90% of the pesticide contamination of the air is by way of these two routes (Lewis and Lee, 1979). Using the example of trifluralin, INRA researchers have shown that the losses by volatilization can reach 10–30% of the quantities applied and that most of these losses to the atmosphere occur during the first few days after application of the product (INRA, 2003).

Factors governing the volatilization process include physicochemicals properties of pesticides, environmental conditions, and agricultural practices.

11.2.3 Physicochemical Properties of Pesticides

The physicochemical properties of the compounds play a key role in determining how they behaves after application. Vapor pressure, water solubility (and thus the Henry's law constant), and adsorption coefficient (soil/water partition) are the most relevant. These characteristics vary by several orders of magnitude from one compound to another: between 10^{-7} and 100 Pa for saturated vapor pressure and between 10^{-3} and 10^3 mg/L for water solubility (Bedos et al., 2002).

Jury et al. (1983) developed a screening model that can outline the importance of the physicochemical properties of pesticides. According to these scientists, the intensity, duration, and time course of the volatilization process from moist soil depends mostly on the Henry's law constant, K_h. On this basis, these authors have defined three categories of pesticide: category I comprises highly volatile pesticides with a nondimensional $K_h \gg 2.65 \times 10^{-5}$; category II comprises moderately volatile pesticides with intermediate values of K_h; and category III, comprises slightly volatile pesticides with $K_h \ll 2.65 \times 10^{-5}$. For pesticides in category I (with a high K_h), the volatilization rate is highest just after application, and then decreases at a rate dependent on whether soil water is evaporating or not. Most organochlorine pesticides belong to this category (I). Pattey et al. (1995) found that the volatilization fluxes of trifluralin and triallate (category I) decreased by a factor of 3 within 4 days. Cessna et al. (1995) also reported such behavior for trifluralin in a study lasting 2 days, as did Majewski et al. (1993) for trifluralin and triallate. According to this model, pesticides belonging to category III tend to accumulate on the soil surface as water evaporates from soil, so volatilization increases with time, or slowly declines if water evaporation does not occur.

As a whole, the correlation between physicochemical characteristics and the flux rates may be valid on the first day after application, because of the wide range of the physicochemical characteristic of pesticides. They give an order of magnitude of the flux occurring after application. These correlations are also limited by the fact that the pesticides applied to soil or crops are commercial formulations, whereas vapor pressure, aqueous solubility, and adsorption coefficient are characteristics of the active ingredient. Since little is known about the differences between the physicochemical properties of the applied formulation and those of the active ingredient, volatilization predictions are highly uncertain (Bedos et al., 2002).

11.2.3.1 Environmental Conditions

A higher air temperature tends to favor volatilization from plants and soils, because the vapor pressure of the pesticide over an aqueous solution is exponentially temperature dependent. According to Spencer and Cliath (1990), a 10°C increase in temperature leads to a three- to fourfold increase in vapor pressure for most pesticides.

On the other hand, interactions with soil temperature and humidity may limit this effect. In the case of an application on vegetation, an increase of 5°C can lead to an increase in the volatilization rate in the case of 2,4-D applied to leaves of barley (Breeze et al., 1992). Van den Berg et al. (1995) observed a similar pattern for parathion applied to a potato crop in a volatilization chamber.

Soil temperature, on the other hand, is a key parameter influencing pesticide volatilization, because pesticide physicochemical properties are temperature-dependent. A temperature increase can be expected to enhance volatilization; however, this behavior is limited by soil dryness. Soil drying tends to promote the adsorption of the pesticide onto the soil matrix, thus limiting its availability for transport to the soil surface (Bedos et al., 2002).

Wind speed seems to have the same influence on pesticide volatilization from soil and from plants. Waymann and Rüdel (1995) found that for wind velocities increasing from 0.4 to 1.7 m/s, volatilization from soil increased from 12 to 31% of the application dose for lindane. On the other hand, these scientists found 62% of lindane volatilized from French beans over 24 h at a wind velocity of 2 m/s versus 52% at a velocity of 0.4 m/s.

Air humidity seems to favor volatilization from soil. According to Grass et al. (1994), at relative air humidities of 31, 49, and 78% (in a wind tunnel), the percentage of trifluralin volatilized over the first day was 66, 64, and 96%, respectively. When pesticide is applied to a crop, a low relative humidity will cause the leaves to dry out more quickly than they would at higher air humidity. This favors adsorption of pesticide molecules on the leaf surface (Van den Berg et al., 1995). Higher air humidity therefore promotes both adsorption of the pesticide by plants and its volatilization (Bedos et al., 2002).

According to several authors (Bor et al., 1995; Jurry et al., 1983; Spencer and Cliath, 1990), volatilization tends to increase with increasing soil water content. This increase of volatilization with soil moisture may be a consequence of a decrease in the number of adsorption sites available on the soil matrix or may be due to the fact that these adsorption sites are not accessible, owing to the presence of a water film, and pesticide molecules, which have low water solubility, cannot reach them. On the other hand, water transfer contributes to pesticide transport inside the soil matrix. The dynamics of the volatilization flux is therefore influenced by all the parameters that govern water transfer (Bedos et al., 2002).

11.2.3.2 Agricultural Practices

It is known that tillage modifies soil properties such as the distribution of the moisture, porosity, organic carbon content, and soil temperature. Wienhold and Gish (1994) reported the effect of tillage on volatilization of alachlor and atrazine.

Droplet size governs the competition between absorption by plant surfaces and volatilization: small droplets tend to evaporate more quickly than larger ones, and larger ones tend to be absorbed faster (Barthélémy et al., 1994; Breeze et al., 1992).

The application date may be important with regard to atmosphere conditions. On the other hand, the effect of the application dose on volatilization from soil has been investigated by Waymann and Rüdel (1995) under semicontrolled conditions. The flux expressed in terms of mass per unit area is proportional to the application dose, whereas the flux expressed as a percentage of the application dose does not increase with dose. The interpretation of such results is based upon the hypothesis of the existence of a thin layer at the soil surface through which transport is mediated solely by molecular diffusion.

11.2.4   Entrainment as Dust

Early reports of pesticide residues contained in airborne particulate matter collected in nonagricultural areas led to widespread belief that wind-blown dust contributes to the occurrence and long-distance transport of pesticides in the atmosphere (Tabor, 1965; Risebrough et al., 1968). According to Peirson and Cambray (1967) (from Lewis and Lee, 1979), dust storms are not uncommon in arid and semiarid regions of the USA and are more frequent in other areas of the world. If the soil particles are sufficiently small and are blown to high altitudes, they may be transported for hundreds of kilometers. Adsorbed pesticides may remain somewhat intact on these particles and may thus be transported along with them. Transglobal movement of very fine particulate matter is well documented by radioactive fallout and may take 2–3 weeks.

Risebrough et al. (1968) collected airborne dust on the coast of Barbados during 1965–1966 and found it to contain 1–164 ng/g of chlorinated pesticides. They estimated that these levels corresponded to only 0.013–0.38 pg/m^3 concentrations in the lower atmosphere. The dust had originated on the west coast of Africa, become entrained in the atmosphere, and was transported some 4,500 km by the northeast trade winds.

11.2.5   Evaporation from Water

Oceans, lakes, and rivers are sinks for pesticide residues. On a global basis, evaporation from water has the potential to be an important route of atmospheric entry for pesticides.

The evaporation loss of a pesticide from water may be relatively easy if the compound is of low solubility and has a moderate vapor pressure. Mackay and Leinonen (1975) calculated theoretical losses for several pesticides and PCBs from saturated water solutions. At 25°C, they predicted a half-life of only 3.1 days for DDT in 1 m^3 of placid water with a 1-m^2 exposed surface. Although the vapor pressure of lindane is 2 orders of magnitude higher than that of DDT, the much greater water solubility of the compound gives it predicted half-life of 191 days (Table 11.4). The scientists suggest that the rate of loss would be much higher for fast-flowing, shallow streams and rough seas, where there is considerable mixing and increased surface exposure.

Table 11.4 Evaporative loss of pesticides from water at 25°C (Mackay and Leinonen, 1975)

Pesticide	Solubility[a] (mg/L)	Vapor pressure (mmHg)	Half-life in water[b] (days)
DDT	1.2×10^{-3}	1×10^{-7}	3.1
Aldrin	0.2	6×10^{-6}	7.7
Lindane	7.3	9.4×10^{-6}	191
Dieldrin	0.25	1×10^{-7}	539
Aroclor 1254	1.2×10^{-2}	7.7×10^{-5}	0.4

Water volume 1 m^3; exposed surface 1 m^2
[a]All compounds studied at saturation levels
[b]Time required for 50% of the compound present in water to be lost to the atmosphere

In general, the levels of pesticides in water are much lower than those found in soil. According to Edwards (1973), the mean levels of organochlorine pesticides in rivers in the USA were found to be 0.2–5.0 ng/L during 1965–1969, while soil levels averaged 0.02–1.6 μg/g for agricultural land and 0.3 ng/g to 0.2 μg/g for nonagricultural land.

11.2.6 Manufacturing and Formulating Processes

Possibilities for atmospheric pollution are associated with the manufacture, formulation, and packaging of pesticides. Most pesticides are manufactured in continuous-process, closed systems that are maintained at slightly negative pressures to avoid leakage. However, no matter how well designed a factory may be or how careful its operation, some atmospheric emissions can be expected. Occasionally, large amounts of pesticides are released into the atmosphere from chemical manufacturing processes through accidents or carelessness (Lewis and Lee, 1976).

In 1964, scientists sampled the air in Fort Valley (Georgia, USA), near a pesticide formulation plant. They found that during the spraying season (May–June) the concentration of DDT in air was 0.007 mg/m^3 and during September (after the spraying season) it was 0.004 μg/m^3. They concluded that most of the pesticide in the air during September could be attributed to emissions from the formulation plant (Tabor, 1965).

According to Heggestad (1974), extensive plant damage was noted in 1970 in a residential area in Chattanooga (Tennessee, USA). The cause was atmospheric emissions of benzoic acid herbicides from a manufacturing plant in the vicinity.

During the accident in Seveso (July 10, 1976) about 3,000 kg of chemicals was released into the air. 2,4,5-Trichlorophenol was produced in the factory and the cloud that resulted from the accident contained various by-products of the process, including 2,3,7,8-tetrachlorodibenzo-*para*-dioxin (TCDD), or dioxin, which it is known to produce cancer in experiments with animals. It is also known that 2,4,5-trichlorophenol, used in the manufacture of herbicides (Corliss, 1999; Gnesda, 1997).

According to many, Bhopal, India, is the site of the greatest industrial disaster in history. On the night of December 23, 1984, a dangerous chemical reaction occurred in the Union Carbide factory. A large amount (about 40 tons) of methyl isocyanate escaped into the air, spreading 8 km downwind, over the city of nearly 900,000 inhabitants (Lancet, 1989). Not very much is known about the environmental impacts of the gas leak from the Bhopal plant. The Indian Council of Agricultural Research has issued a preliminary report on damage to crops, vegetables, animals, and fishes due to the accident but this offers few conclusive findings since they were reported in the early stages after the disaster. This report, however, did indicate that the impact of whatever toxic substances emerged from the Union Carbide plant were highly lethal to exposed animals. Plant life was also severely damaged by exposure to the gas. There was also widespread defoliation of trees, especially in low-lying areas (Morehouse and Subramaniam, 1986).

11.2.7   *Disposal Processes*

The disposal of pesticides can often lead to airborne emissions of pesticide residues trough volatilization or can serve as a source of other air pollutants that may be formed during the disposal process.

Most pesticides will undergo some degree of degradation by hydrolysis or metabolism by microorganisms in the soil, but will often have alteration products of equal or greater toxicities and/or volatilities. Even stable pesticides that are disposed of through burial can remain a source of air pollution for many years unless they are buried in impervious containers (Lewis and Lee, 1976).

Kennedy et al. (1972) analyzed the major volatile products formed from the incineration of 20 common pesticide formulations at 900°C. Table 11.5 summarizes their findings. Smith and Ledbetter (1971) defined the toxic hazard of fires involving organophosphorus pesticides. They concluded that the reactions of these chemicals, which seem likely to take place in a fire, could be classified as decomposition, isomerization, and polymerization reactions. Isomerization reactions were thought to produce products which had toxicity comparable to that of the original compound. Solutions of malathion and parathion in a variety of solvents were ignited. The gases above the fire and the residues were analyzed. It was found that virtually all the pesticide that passed through the flame was destroyed, although dimethyl fumarate was positively identified as a decomposition product.

11.3   Pesticides and Soil

Soil is a dynamic, living matrix that is an essential part of the terrestrial ecosystem. It is a critical resource not only for agricultural production but also for the maintenance of most life processes. The living portion of the soil is composed of plant roots as well as communities of biota (living soil organisms) critical to the function of soils. Soils have enormous numbers of diverse organisms assembled in complex

Table 11.5 Volatile products from burning of analytical-grade pesticides at 900°C (Kennedy et al., 1972)

Pesticide	Combustion products identified
Atrazine	CO, CO_2, Cl_2, HCl, NH_3
Bromacil	CO, CO_2
Carbaryl	CO, CO_2, NH_3, O_2
Dalapon	CO, CO_2, Cl_2, O_2, HCl
DDT	CO, CO_2, Cl_2, HCl
Dicamba	CO, CO_2, Cl_2, HCl, O_2
Dieldrin	CO, CO_2, Cl_2, HCl, O_2
Diuron	CO, CO_2, Cl_2, NO, HCl, NH_3
Dinoseb	CO, CO_2, NH_3
DSMA	CO, CO_2, O_2
Malathion	CO, CO_2, SO_2, H_2S, O_2
Dibromochloropropane	CO, CO_2, Cl_2, HCl
Paraquat	CO, CO_2, NH_3
Picloram	CO, CO_2, NH_3, Cl_2
PMA	CO, CO_2
Trifluralin	CO, CO_2, NH_3
2,4-D	CO, CO_2, Cl_2, HCl, O_2
2,4,5-T	CO, CO_2, Cl_2, HCl, O_2
Vernolate	CO, CO_2, NH_3, O_2
Zineb	CO, CO_2, H_2S, NH_3

and varied communities. Soil that contains a balance of active biological components is essential to all agricultural systems (DIPNR, 2004).

Pesticides are widely used to control pests that affect agricultural crops. Usually, most of the applied pesticide comes into contact with soil, which is a major "incubation" chamber for the decomposition of any organic waste (pesticides, solid wastes, sewage, etc.). Pesticides in the soil do not remain stationary, but they move within the soil. They move by both diffusion throughout the medium in which they are dissolved and mass flow of the water in the soil.

11.3.1   Fate of Pesticides in Soil

The fate of a pesticide applied to soil depends largely on two of its properties: persistence and solubility.

Persistence is the ability of a pesticide to resist breakdown (degradation) into compounds that have different chemical structures and properties (WSU, 1996). It defines the "lasting power" of a pesticide. Persistence time, which is related to the rate of disappearance, varies from days to years depending on the type of pesticide, soil moisture, organic matter, temperature, and pH. In general, nonpersistent pesticides disappear from soil in less than 1 month, while moderately persistent chemicals take from 1 to 3 months. Persistent pesticides are present for many months (years in some cases) after application (Doxtader and Croissant, 1992).

Some pesticides persist in the soil but do not build up in animals. Atrazine, for example, does not bioaccumulate, although it is a persistent pesticide.

Persistence is expressed as a "half-life." Each half-life measures the amount of time it takes for half of the original amount of pesticide in the soil to be deactivated, or lose its ability to work effectively. However, half-life can also be defined as the time required for half the amount of the applied pesticide to be completely degraded, or broken down (Bicki, 1989).

Probably, the single most important property influencing a pesticide's movement with water is its *solubility*, or ability to dissolve in water. When a pesticide enters soil, some of it will stick to soil particles, particularly organic matter, through the process of adsorption and some of it will dissolve and mix in the water between soil particles, called "soil water."

11.3.1.1   Chemical Degradation

Chemical degradation is the breakdown of a pesticide by processes not involving a living organism. Most pesticides are chemically stable across a broad range of conditions. A few of them degrade because of their lack of stability in soil. For example, the herbicides treflan and atrazine are subject to chemical decomposition (Doxtader and Croissant, 1992).

11.3.1.2   Photodegradation

The process of photodegradation, or photochemical degradation, involves the chemical transformation of substances under the action of sunlight. Like other degradation processes, photodegradation reduces the amount of chemical present and lowers the level of pest control. Photochemical reactions are limited to the soil surface. Mechanical incorporation into the soil during or after application, or by irrigation water or rainfall following application, can reduce pesticide exposure to sunlight. Pesticides susceptible to photodegradation include 2,4-D, treflan, banvel (dicamba), and parathion (Doxtader and Croissant, 1992; PMEP, 2003).

11.3.1.3   Degradation by Soil Microorganisms

Some pesticides are destroyed in soil by some species of microorganisms. Soil bacteria and fungi can decompose the majority of these compounds. This occurs when microorganisms use a pesticide as food. Under proper soil conditions, i.e., favorable temperature and pH levels, adequate soil moisture, aeration, and fertility, the microbial degradation can be rapid and thorough. If soil is cold (below 50°F) or dry, microbial activity is slow and degradation is limited. Some microorganisms function best under aerobic conditions, while others are most active in the absence of oxygen. Microbial abundance and activity are stimulated by the addition of organic materials to soil, which usually results in faster rates of pesticide degradation. Some pesticides bond to the humus colloids and clay minerals. This process limits the access of microorganisms to the pesticides and slows the rate of decomposition. Generally, microbial degradation occurs most rapidly in warm, moist soil that contains readily decomposable organic matter such as manure or fresh plant material. The rate at which the degradation occurs depends also on the chemical nature of the pesticide and on the how the soil has been treated in the past with the particular pesticide. Those pesticides that are decomposed slowly or incompletely by microorganisms, such as the herbicide tordon (picloram), show long persistence times. On the other hand, 2,4-D disappears quickly because it is degraded by a variety of common soil microorganisms (Doxtader and Croissant, 1992; INRA, 2000; PMEP, 2003).

11.3.1.4   Leaching

Leaching is the movement of pollutants through the soil by rain or irrigation water, as the water moves downward through the soil.

There are several factors that influence pesticide leaching. Soil texture and structure, for example, influence the depth and the rate of chemical leaching. It is known that sandy and gravelly soils allow water and pesticides to leach through quickly. A heavy clay soil does not allow for rapid leaching. Water that moves down through soil or through cracks and worm tunnels transports water-soluble substances. The herbicides 2,4-D and tordon are leached easily in soil. Chemicals such as paraquat which are strongly absorbed onto clay and humus show limited downward movement. The leaching of pesticides can contaminate groundwater (Doxtader and Croissant, 1992; PMEP, 2003).

11.3.1.5   Adsorption–Desorption

Adsorption is a process in which the pesticide forms chemical bonds with colloidal materials, such as soil organic matter and clay particles. Adsorption is an extremely important process affecting the fate of pesticides. Strongly adsorbed pesticides will be less mobile when leached through soil than are weakly adsorbed particles (WSU, 1996).

Pesticide adsorption in soil depends on the pesticide properties, soil moisture content, soil acidity (pH), and soil texture. However, soil-adsorbed pesticides can be lost by erosion (PMEP, 2003).

One of the most useful indicators for quantifying pesticide adsorption in soils is the "partition coefficient," or PC value. The PC value is defined as the ratio of the amount of pesticide bound to soil particles to the amount of pesticide dissolved in the soil water. For a given amount of pesticides, those with small PC values are more likely to be leached than those with large PC values (Rao et al., 1998).

Desorption is the converse process of adsorption, i.e., the decrease in the amount of substance adsorbed.

11.3.1.6   Volatilization

See Sect. 11.2.2.

11.3.1.7   Uptake by Higher Organisms

Some pesticides must be taken up by plants, insects, or other target organisms to be effective. Pesticides also are taken up by nontarget organisms through root or leaf absorption by plants and ingestion or contact by animals. Once inside the tissue of an organism, they are transformed by the action of enzymes or stored unchanged. Pesticide degradation within the plant is an important mechanism of detoxification; it helps reduce the hazard of residues in food (Doxtader and Croissant, 1992).

11.3.1.8   Aerial Movement

The movement from soil to the atmosphere occurs under many conditions. Pesticides move from soil into the atmosphere by volatilization and wind erosion. It is known that not all pesticides are subject to volatilization, but for a few, this is a major pathway of loss from soil.

11.4   Pesticides and Water

The main purposes of water for human beings are for public water supply, aesthetics and recreation, agriculture, industry, and aquatic life.

Unlike air pollution, where the leading emission factors are natural, water pollution is mostly caused by human involvement. Among other anthropogenic water pollutants, pesticides are a serious source of pollution.

There are three basic ways that applied pesticides may reach surface and groundwater, i.e., runoff, run-in, and leaching (Bicki, 1989):

1. *Runoff* is the physical transport of pollutants over the soil surface by rain or irrigation water that does not soak into the soil. Pesticide runoff can occur either with the pesticide as a solute or as the pesticide adsorbed to soil particles (soil erosion).
2. *Run-in* is the physical transport of pollutants directly to groundwater. For example, this can occur in areas of karst carbonate (limestone) aquifers, which contain sinkholes and porous or fractured bedrock. Rain or irrigation water can carry pesticides through sinkholes or fractured bedrock directly into groundwater.
3. *Leaching* (see Sect. 11.3.1.4).

The greatest potential for pollution effects of pesticides is through contamination of the hydrological system. Water is one of the primary pathways by which pesticides are transported from the site of their first application to other areas of the environment.

11.4.1 Pesticides and Surface Water

The surface water pollution by pesticides can be divided in two main categories, i.e., intentional and unintentional.

Intentional includes the direct application of pesticides to water to control harmful insects or waterweeds. *Unintentional* pollution of water occurs in some cases, e.g., during the spraying of crops near streams, lakes, dishes, etc. In some cases, some quantities of pesticide residues flow through the drainage water or are a result of runoff in drainage ditches, rivulets, rivers, and drainage basins. Also, pesticides frequently volatilize into the air. Depending on their stability they can be taken up into the atmosphere and transported over long distances. They may then return to the ground through rainfall, and through the soil into water basins. Pesticides are found in rain in parts of Europe, sometimes at levels so high that it would be illegal to supply it as drinking water. Rainwater has been found to contain lindane in Spain and 2,4-D in Italy. Until relatively recently, pesticides were regularly found in drinking water. The European Union set the limit on a precautionary basis at 0.1 mg/L. In 1992, 3% of samples contained a pesticide in excess of this limit; in 1998, only 0.01% of samples contravened the limit, indicating how successful – at a cost – water companies have been in reducing pesticides in the water supply (PAN UK, 2000).

Many pesticides are detected in different surface waters. They can be transferred over great distances. Pesticides can be traced for hundreds of kilometers in rivers either attached to soil particulates or in solution. Among the synthetic organic pesticides, chlorinated hydrocarbons are the most dominant of these pollutants, because of their persistence. Because of the great length of time, these hydrocarbons can invade areas of the environment they were not meant for and spread into soils, runoffs, and water environments. Their toxic effects on living organisms pervade fatty membranes around nerves and disrupt the movement of ions between the fibers (Buchanan and Horwitz, 1997).

In Greece, some quantities of chlorinated hydrocarbons and of their metabolites were found in the water of Salonica Bay and in the body of some mussels during 1978–1979. Additionally, in the drainage basin of the Pinions river about 1,800 tons of pesticides is rejected per year, and 200 tons of these compounds ends up in Pagasiticos Bay (Chrysoyelos, 1988).

Organophosphorus insecticides and carbamates, and generally the new generation of compounds will not create side effects similar to those caused by chlorinated hydrocarbons, because they do not have a tendency for bioaccumulation or biological magnification, but are degraded in a short time. Fenitrothion, for example, is degraded in a period of 2–4 days in lakes and marshes. Also, the pyrethrin deltamethrin has an active life of 1 h in water (Balayannis, 1988).

11.4.2 Pesticides and Groundwater

Groundwater is a region within the earth that is wholly saturated with water. It is a precious natural resource for public and agricultural communities. The toxic qualities of pesticides used in agriculture and that make them effective for pest control create a potential for groundwater contamination.

Pesticides applied correctly to a site may be moved downward with rain or irrigation water, reaching the water table below. This type of contamination is called *non-point-source pollution*. Pesticides may enter a well directly from spillage or back-siphoning, thus entering the groundwater directly. This is called *point source pollution*. Because groundwater moves slowly, contaminants do not spread quickly. After pesticides reach groundwater, they may continue to break down, but at a much slower rate because of less available light, heat, and oxygen. Thus, they can remain underground in slow-moving plumes for an indefinite period. When groundwater becomes contaminated, the polluted water may eventually appear in the surface water of streams, rivers, and lakes. Because of the complex nature of groundwater, when the contamination is detected it is often widespread. Even if the contamination is stopped, it may take years before an aquifer can purify itself through natural processes (PMEP, 2003).

Three major factors determine whether a pesticide is likely to reach groundwater. These are (1) pesticide properties (persistence and adsorption), (2) soil properties (permeability, organic matter, soil texture, structure, and moisture), and (3) site conditions (including rainfall, irrigation, and depth to groundwater) (Buttler et al., 1998; WSU, 1996).

11.5 Pesticides and Biodiversity

Biodiversity is the variety of existing life (plants, animals, and microorganisms) in the surrounding environment; it takes into account the number of species present, the amount of genetic variation within a species, and the habitats in which the species live.

Pesticides can adversely affect crops, flora, fauna, waterways, and dwellings. It is estimated that 70–90% of ground-applied pesticides and only 25–50% of aerially applied pesticides actually reach their target area (WWF, 1999). The remaining quantity of chemicals has the potential to impact nontarget organisms (vertebrates, invertebrates, microorganisms, and plants), and to become widely dispersed in the environment. On the other hand, pesticides can "drift" from the site of application in numerous ways (transport of airborne particles via air currents, runoff from fields into surface waters, leaching through soils into groundwater, transport by ocean currents carrying the pesticides between continents, etc.). All these mechanisms together with the ability of some chemicals to persist in the environment for a number of years means that pesticides have the potential to affect an immense variety of wildlife, from birds feeding on infested seeds to polar bears feeding on saltwater fish, hundreds of kilometers away (Oliver, 2001).

In relation to the potential impact of pesticides on nontarget organisms, the mode of action is the most significant parameter. The drawback of this approach is that the details of the mode of action are unknown but nevertheless still identify the potential area of concern for environmental impact on nontarget organisms. In some cases the mode of action is completely unknown (Morley, 1999). The current thoughts on mode of action of pesticides are summarized in Table 11.6 (Coats, 1982; Corbett et al., 1984; Hassall, 1982).

11.5.1   Direct Lethal Effects

It has been shown by several researchers that some pesticides are *acutely toxic* in animals. For example, carbamate insecticides that have a neurotoxic effect on insects are also known to affect birds, fishes, bees, and mammals (Oliver, 2001).

Boutin et al. (1999), in their experiments in southern Ontario, found that birds ingest granular pesticides as grit or as mistaken food items. Usually, many birds are particularly vulnerable because they live and feed in and around farms. This is especially true for birds in apple orchards, because of the toxicity of the pesticides used and the number of applications, i.e., up to ten to 12 per season. Most carbamates and organophosphates used in orchards are extremely toxic to birds, with lethal doses ranging from a few milligrams per kilogram to less than 1 mg/kg.

Mass deaths of gulls in southwestern Greece during 1989 were attributed to pesticides applied in nearby agricultural areas (Roumpos et al., 2001). Some cases of deaths of the European brown hare (*Lepus europaeus*) were noted in England

Table 11.6 Illustrations of pesticides and their sites of action (Coats, 1982; Corbett et al., 1984; Hassall, 1982)

Site of action	Pesticides
Respiration	Arsenical, copper compounds, dinitroaniline herbicides (secondary site of action), oxathiin carboxanilides, dinitrophenols, pentachlorophenol, hydroxybenzonitriles, trisubstituted, rotenone, phosphine, hydrogen cyanide
Nervous system	Organophosphorus compounds, pyrethroids, N-methyl carbamates or $N,N,$-dimethyl carbamates, (most) organochlorine compounds, nicotine, avermectins, chlorodimeform and related compounds
Biosynthesis	Acylalanines, hymexazole, cycloheximidae, pyridazinones, triazoles, aminotriazole, thiocarbamates, imidazoles, pyrimidines, dichlobenil, diflubenzuron, glyphosate, ethirimol, tricyclazole
Photosynthesis	Herbicides: straight-chain, substituted, and cyclic ureas, triazines, acylanilides, phenylcarbamates, triazinones, phenolic herbicides, nitrodiphenyl ethers
Cell growth and development	Benzimidazoles and related compounds, dicarboxamides, N-phenylcarbamates, dinitroanilines, phosphoramidates, sulfonylureas, maleic hydrazide, juvenile hormones and analogues and precocenes
Nonspecific	Mercury compounds, sodium fluoride, fungicides of captantype, petroleum, and tar oils, long-chain guanidino fungicides, chloracetanilides, chlorinated short-chain aliphatic carboxylates, chlorthalonil, alkyl bisdithiocarbamates

and in France as well on agricultural land that had been sprayed with the herbicide paraquat (Edwards et al., 2000).

Pesticides possibly are a direct threat to nontarget insect species, including some beneficial species. For instance, according to Kearns and Inouye (1997), the broad-spectrum insecticides used to control some grasshopper populations on range lands in the USA killed many other insects, including native pollinators. On the other hand, in New Brunswick, populations of native bees and syrphid flies were adversely affected when some forests were sprayed to control spruce budworms. Similar observations in Greece during last 55 years have shown side effects of insecticides on useful species of insects (Tzanakakis, 1961; Alexandrakis, 1979, 1989; Alexandrakis and Paraskakis, 1989; Zouridakis et al., 1995; Roumpos et al., 2001), and in Cyprus (Orphanidis, 1993) (for details, see Sect. 11.3).

Encapsulated methyl parathion caused massive bee kills in some cases, because foraging bees took the capsules back to their hive and stored them together with pollen. Thus, foraging bees were killed transporting and collecting contaminated pollen. On the other hand, young hive bees were killed by the poisoned food (Atkins et al., 1978; Johansen, 1977).

11.5.2 Indirect, Large-Scale Effects

Use of pesticides to control pathogens and pests leads to the development of a long list of serious indirect problems. An indirect effect of a treatment is to kill nontarget organisms since by their nature most pesticides are broad-spectrum toxicants designed to kill living organisms. Many indirect effects are not measurable as acute toxicity.

The use of herbicides may have indirect and drastic effects on biodiversity over a very large scale. The decrease of plant diversity begins a cascade of events which ultimately affects a multitude of species (Kearns and Inouye, 1977). For instance, increases in chick mortality rates have been caused by increase in herbicide use (Sotherton et al., 1988).

According to Kearns and Inouye (1997), the extensive use of herbicides reduces the abundance of many plants, which subsequently decreases the diversity of insects and pollinators that depend on them. Since thousands of wild plants depend on insect pollinators to produce fruits and seeds, this leads to a further decrease of plant diversity. This vicious circle is intensified by the use of broad-spectrum pesticides. So, a decrease in plant diversity from the use of herbicides affects not only organisms that consume plants, fruits, or seeds, or that use plant materials for nest construction, but also those that rely on insects as a food source. Partridge and pheasant chicks between 2 and 3 weeks of age rely on insects as the protein source for growth. An increase in the use of herbicides has been shown to be responsible for increased chick mortality due to lack of food (Hill, 1985; Sotherton et al., 1988).

Studies by Haughton et al. (1999) in England have shown that herbicide glyphosate drift at rates of more than 360 g active ingredients/ha into arable field margins could result in significant losses of important arthropod predators in farmland and a reduction in spider biodiversity in agroecosystems.

A significant indirect intervention of pesticides in agroecosystems is through food chains. Because pesticides often travel from one level in the food chain up to the next, they can have damaging effects on many species that never came into direct contact with these chemicals.

Pesticides with bioaccumulation in the food chain decimate or cause the disappearance many species of organisms. For instance, relevant research conducted in Messara in southeast Crete by Greek ornithologists in 1976 showed that some species of entomophagous birds had disappeared entirely from the area, owing to side effects from air sprays against the olive fly. Some of these species of birds were the Eleonora falcon (*Falco eleonorae*), the Eurasian kestrel (*F. tinnunculus*), the European scops owl (*Otus scops*), the barn swallow (*Hirundo rustica*), and the common nightingale (*Luscinia megarhynchos*). In contrast, the common buzzard (*Buteo buteo*), which feeds on rodents and snakes, did not show any variation in its population (Kalopissis, 1981).

Pesticides are generally considered the main cause of the appearance of toxicity in animals through food chains. Relevant research in northern Greece during 1990–1995 that included 926 samples of several animals (sheep, birds, cats, dogs, and bees) showed that 78% of the cases of toxicity were caused by pesticides, while the remaining 22% were from other causes. Similar research by Tsoukali-Papadopoulou during 1996 showed a significant increase of the number of poisonings of productive animals (cattle, sheep, goats) and birds in 1986; carbamates outnumbered among other kinds of pesticides in cases of poisoning (Roumpos et al., 2001).

During a decade of research (1990–1999) by Antoniou et al. (2001) in several areas of Greece (Thessaly, Epirus, and Macedonia) including 902 samples from some animals (cattle, sheep, goats, birds, etc.) and 302 sample of forage, baits, and other materials, toxicants were detected in 34.04% of the biological samples which caused the death of the animals in a percentage of about 80%. Dominant among these toxicants were pesticides; carbamates had the greatest frequency of appearance (Table 11.7). On the other hand, toxicants were detected at a level of 38.74% in forage, baits, and other materials. Of interest was the great percentage of positive samples (50%) in wild birds and bees (29.82%) as well. Some pesticides (such as organochlorines) are biomagnified, which describes the increasing concentrations of a chemical with higher trophic levels.

Organisms at the bottom of the food chain tend to bring into their bodies a higher concentration of the pollutant than is found in the surrounding soil or water. Since the consumer at any trophic level needs to consume a lot of biomass from the lower trophic level, the consumer will ingest large quantities of the pollutant. Some pesticides such as DDT are fat-soluble and will be accumulated in the fat stores of the consumer, where it will remain "available" for the consumers at the next trophic level (Oliver, 2001).

Table 11.7 Toxicants detected in several species of animals during the decade 1990–1999 (Antoniou et al., 2001)

Species	Number of samples	Positives							Total positives	Distribution (%)
		Organophosphates	Carbamates	Herbicides	Rodenticides	Alkaloids	Heavy metals	Miscellaneous		
Horses	11	–	1	–	–	1	–	–	2	18.18
Cattle	92	1	3	–	–	–	2	–	6	6.52
Sheep	187	7	23	2	2	–	32	–	66	35.29
Goats	95	8	6	2	–	1	1	–	18	18.95
Swine	40	3	1	–	2	1	–	–	7	17.50
Domestic birds	52	5	11	–	–	–	–	–	16	30.77
Wild birds	28	2	10	–	–	1	–	1	14	18.18
Bees	57	8	6	3	–	–	–	–	17	29.82
Dogs, cats	318	39	96	3	3	15	–	1	157	49.37
Animals for fur	22	1	3	–	–	–	–	–	4	18.18
Total	902	74	160	10	7	19	35	2	307	34.04

Table 11.8 Biomagnification of DDT in different trophic levels (Kotsovinos, 1985)

Species	Trophic level	DDT (ppm)
Laminaria (seaweed)	1	0.001
Echinus (herbivore)	2	0.024
Lobster	3	0.027
Liver of a fish	4	1.560

Low levels of DDT (1 ppb) were found in the water of a marine area. The same level of this pesticide was detected in the body of *Laminaria* (a species of sea-weeds), but the level of DDT in the liver of a species of fish from the same marine area was 1,560 ppm) (Kotsovinos, 1985) (Table 11.8).

11.5.3 Disruption of the Ecological Balance

In agricultural practice, many people believe that more pesticides mean fewer pests and higher crop yields. According to Kegley et al. (1999), even though the use of insecticides increased tenfold from 1945 to 1989 in California, crop losses from insects nearly doubled in the same time period, from 7 to 13%.

Pesticides have the potential to kill beneficial organisms which serve important functions such as controlling pests, providing nutrients to the plants, and aerating the soil. This process disrupts the natural ecological balance between the beneficial and the destructive organisms and encourages more pest and disease attack. The result is the need for even more chemicals, beginning a vicious circle of dependence.

Pesticide residues in soil, in addition to eliminating or reducing parasitic microbes, are also toxic to the nonparasitic and ecologically useful soil microbial and vertebrate population. Pesticides may reduce certain microorganism populations, while they stimulate the growth of others, especially the saprophytic and spore-forming types. These processes may disrupt the ecological balance in the soil microenvironment, first by simplifying the microbial population, and possibly by reducing soil fertility and its ability to support life. Nontarget or residual pesticide toxicity would also disrupt the population of some of the valuable soil invertebrates such as earthworms, predatory mites, centipedes, and carabid beetles (Iyaniwura, 1991).

According to Edwards (1965), the most usual effect of agricultural practice is to decrease the number of species of soil organisms; the few species that remain are often able to multiply rapidly until the total numbers are greater than they were originally. The main difference from cultivation is that the effect of chemicals lasts longer. Whereas plowing or rotation may only change the balance of the soil fauna or flora for a matter of weeks or at most months, persistent chemicals can alter it for months or years.

Having ground cultivations, the disruption of ecological balance due to the use of pesticides is not a rare occurrence. For instance, an increase of the population density of some apple mite species was observed in 1950 in the area of Naoussa (northern Greece), after the chronic use of the insecticide DDT to control the codling moth *Carpocapsa pomonella* on apple orchards, and in 1955 in the area of Leonidion (southern Greece) as well after the use of the same chemical on pear orchards. These findings were attributed to the killing of natural parasites of the mites during the chemical control of the codling moth (Tzanakakis, 1961).

Bartlet (1964) reported population outbreaks of some secondary entomological species after destruction of their entomophages, placing them in a significant position among the primary harmful crop species.

In Crete, the increase of population density of the white scale *Aspidiotus nerii* in the local olive orchards was attributed to the reduction of populations of its hymenopterous parasite *Aphytis chilensis*, as a result of air and ground sprays with organophosphorus insecticides to control the olive fly *Bactrocera oleae* (Alexandrakis, 1979, 1989). The increase of the population density of the olive black scale *Saissetia oleae* was also attributed to the elimination of its parasites by the same cause (Alexandrakis and Paraskakis, 1989; Paraskakis, 1985; Zouridakis et al., 1995). Also in Cyprus, the population outbreaks of *S. oleae* were attributed to the elimination of its parasites owing to the air-bait sprays against the olive fly (Orphanidis, 1993).

References

Akesson NB (1973) Application equipment for pesticide spraying. Workshop on Pesticide Spray Technology, Oakland, CA.

Akesson NB and Yates WE (1964) Problems relating to application of agricultural chemicals and resulting drift residues. Ann. Rev. Entomol. (9): 285.

Alexandrakis V (1979) Contribution a l' étude d' *Aspidiotus nerii* BOUCHE (Homoptera, Diaspididae) en Crète. Thèse Univ., Bordeaux.

Alexandrakis V (1989) Effect of *Dacus oleae* control sprays by air or ground on the ecology of *Aspidiotus nerii* Bouché (*Horn. Diaspididae*). International Symposium of Olive Growing, Cordoba, Spain, Sept 1989.

Alexandrakis V and Paraskakis M (1989) The side effects of insecticides on useful organisms. Proceedings of Symposium: "Agrochemical and Environment," Chania, Dec 1989 (in Greek).

Antoniou V, Zantopoulos N, and Tsoukali E (2001) http://kynigos.net.gr/diaxirisi/articles/dilitiria-seis.html-12k

Atkins EL, Kellum D, and Atkins KW (1978) Encapsulated methyl parathion formulation is highly hazardous to honeybees. Am. Bee J. (118): 483–485.

Balayannis PG (1988) The presence of pesticides into superficial and underground waters. Proceedings of the International Symposium: "Pesticides: Problems and Alternative Solutions," Athens, Sept 1988 (in Greek).

Barthélémy P, Bouvier JL, and Jouy L (1994) Technique de pulvérisation: Comment limiter la dérive? Perspect. Agric. (194): 79–87.

Bartlet BR (1964) Integration chemical and biological control. In: Bach De (Ed.) *Biological Control of Insect Pest*. Chapman and Hall, London.

Bedos C, Cellier P, Calvet R, Barriuso E, and Gabrielle B (2002) Mass transfer of pesticides into the atmosphere by volatilization from soils and plants: overview. Agronomie (22): 21–33.

Bicki TJ (1989) Pesticides and groundwater. Pesticides as potential pollutants. Land and Water No 12.

Bor G, Van den Berg F, Smelt JH, Smidt RA, Van de Peppel-Groen AE, and Leistra M (1995) *Volatilization of Triallate, Ethoprophos and Parathion Measured with Four Methods After Spraying on a Sandy Soil*. SC-DLO, Agricultural Research Department, Winand Staring Centre for Integrated Land, Soil and Water Research.

Boutin C, Freemark KE, and Kirk DA (1999) Farmland birds in Southern Ontario: field use, activity patterns and vulnerability to pesticide use. Agric. Ecosyst. Environ. (72): 239–254.

Breeze VG, Simmons JC, and Roberts MO (1992) Evaporation and uptake of the herbicide 2,4-D-butyl applied to barley leaves. Pestic. Sci. (36): 101–107.

Buchanan M and Horwitz C (1997) http://www.umich.edu/~gs265/society/pollution.htm

Buttler T, Martinkovic W and Nesheim ON (1998) http://edis.ifas.ufl.edu/BODY_PI002.

Cessna AJ, Kerr LA, Pattey E, Zhu T, and Desjardins RI (1995) Field comparison of polyurethane foam plugs and mini-tubes containing Tenax-Ta resin as trapping media for the aerodynamic gradient measurement of trifluralin vapour fluxes. J. Chromatogr. A (710): 251–257.

Chrysoyelos N (1988) Pesticides: their invasion and dominion in the Greek area. Proceedings of the International Symposium: "Pesticides: Problems and Alternative Solutions." Athens, Sept1988 (in Greek).

Coats JR (Ed.) (1982) *Insecticide Mode of Action*. Academic, New York.

Corbett JR, Wright K and Baillie AC (1984) *The Biochemical Mode of Action of Pesticides*(2nd edn). Academic, New York.

Corliss M (1999) http://www.getipm.com/articles/seveso-italy.htm

Cunningha RT, Brann JL Jr. and Fleming GR (1962) Factors affecting the evaporation of water from droplets in airblast sprayings. J. Econ. Entomol. (55):192.

Decker GC, Weinham CJ and Bann JM (1950) A preliminary report on the rate of insecticide loss from treated plants. J. Econ. Entomol. (43): 919.

DIPNR (2004) http://www.dlwc.nsw.gov.au/care/soil/3funct.htm

Doxtader KG and Croissant RL (1992) http://www.oda.state.or.us/nrd/water_quality/BMPs/Pesticides/Fate_of_Pesticides.pdf

Durham WF (1979) Human health hazards of respiratory exposure to pesticides. In: *Air Pollution from Pesticides and Agricultural Processes*, CRC, Boca Raton, FL.

Edwards CA (1965) Some side-effects resulting from the use of persistent insecticides. Ann. Appl. Biol. (55): 329–331.

Edwards CA (1973) Pesticide residues in soil and water. In: *Environmental Pollution by Pesticides*, Edwards CA ed., Plenum, London and New York.

Edwards PJ, Fletcher MR, and Berny P (2000) Review of the factors affecting the decline of the European brown hare, *Lepus europaeus* and the use of wildlife incident data to evaluate the significance of paraquat. Agric. Ecosyst. Environ. (79): 95–103.

Furmidge GGL (1963) The application of flying-spot scanning to particle size analysis in the formulation of pesticides. Analyst (88): 686

Grass B, Wenclawiak BW and Rüdel H, Influence of air velocity, air temperature, and air humidity on the volatilization of trifluralin from soil. *Chemosphere* 28 (1994), pp. 491–499.

Gnesda D (1997) Dioxin exposure and cancer risk: a 15-year mortality study after the "Seveso accident." Epidemiology (6): 8.

Hassall KA (1982) *The Chemistry of Pesticides*. Macmillan, New York.

Haughton A, Bell JR, Boatman ND and Wilcox A (1999) The effects of different rates of the herbicide glyphosate on spiders in arable field margins. J. Arachol. (27): 249–254.

Heggestad HE (1974) Air pollutants from and effects on agriculture. Presented at the APCA Conference on Control Technology for Agricultural Air Pollutants, Memphis, Tennessee, March 1974.

Hill DA (1985) The feeding ecology and survival of pheasant chicks on arable farmland. J. Appl. Ecol. (22): 645–654.

Hurtig H (1973) Chemicals in the environment. Some aspects of agricultural chemicals. In: *Environmental Quality and Safety*, vol. 2, Coulston F and Korte Feds., Academic, New York.

INRA (2000) http://www.inra.fr/PRESSE/mars00/nb1.htm

INRA (2003) http://www.inra.fr/PRESSE/janv03/gb/nbl.htm

Iyaniwura TT (1991) Non-target and environmental hazards of pesticides. Rev. Environ. Health 9 (3): 161–176.

James JA and Vaughan LM (1970) Measuring particle drifts to four miles. J. Appl. Meteorol. (9): 79.

Johansen CA (1977) Pesticides and pollinators. Ann. Rev. Entomol. (1): 51–54.

Jurry WA, Spencer WF, Farmer WJ (1983) Behavior assessment model for trace organics in soil. I. Model description. J. Environ. Qual. (12): 558–564.

Kalopissis ITh (1981) *Air Sprays Against Olive Fly and Nature's Protection* (in Greek).

Kearns CA and Inouye DW (1997) Pollinators, flowering plants, and conservation biology: much remains to be learned about pollinators and plants. BioScience (47): 297–307.

Kegley S, Neumeister L and Martin T (1999) http://www.panna.org/resources/documents/disruptingSum.dv.html

Kennedy MV, Stojanovic BJ and Shuman FL Jr (1972) Analysis of decomposition products of pesticides. J. Agric. Food. Chem. (20): 341.

Kiigemagi U and Terriere LC (1971) Losses of organophosphorus insecticides during application to the soil. Bull. Environ. Contam. Toxicol. (6): 336.

Kotsovinos NE (1985) *Pollution and Protective of the Environment*, "Plaisio" (ed.), Athens, (in Greek).

Koutselinis A (1997) *Toxicology*, vol. I, Parissianos, Athens.

Lancet (1989) *"Round The World: India – Long Term Effects of MIC."* Issue 644.

Lee RE Jr (1979) *Air Pollution from Pesticides and Agricultural Processes*, CRC, Boca Raton, FL.

Lewis RG and Lee RE Jr (1979) Air pollution from pesticides: sources, occurrence and dispersion. In: *Air Pollution from Pesticides and Agricultural Processes*, CRC Press Inc. ed., USA.

Mackay D and Leinonen PJ (1975) Rate of evaporation of low-solubility contaminants from water bodies to atmosphere. Environ. Sci. Technol. (9): 1178.

Majewski MS and Capel PD (1995) *Pesticides in the Atmosphere: Distribution, Trends, and Covering Factors*, Ann Arbor, Chelsea, MI.

Majewski M, Desjardins R, Rochette P, Pattey E, Selber J, and Glotfelty D (1993) Field comparison of an eddy accumulation and an aerodynamic-gradient system for measuring pesticide volatilization fluxes. Environ. Sci. Technol. (27): 121–128.

Morehouse W and Subramaniam MA (1986) *The Bhopal Tragedy: What Really Happened and What It Means for American Workers and Communities. At Risk.* Council on International and Public Affairs.

Morley HV (1999) www.iscu-scope.org/downloadpubs/scope49/chapter06.html

OECD (2000) *Environmental Performance Reviews – Greece.*

Oliver J (2001) http://www.dal.ca/~dp/reports/zoliner/oliverst.html

Orphanidis PS (1968) *Agricultural Pharmacology.* Part I, Spyrou (ed.), Athens (in Greek).

Orphanidis G (1993) Integrated pest management in Cyprus: implementation and perspectives. Proceedings of 5th Pan-Hellenic Entomological Conference, Athens, Nov 8–10, 1993 (in Greek).

PAN UK (2000) http://www.pan-uk.org/pestnews/pn49/pn49p5.htm

Paraskakis M (1985) Evaluation of parasites of the olive black scale in Crete. 1st Scientific Meeting for Tree Crops. Chania, Nov 1985 (in Greek)

Pattey E, Cessna AJ, Desjardins RL, Kerr LA, Rochette P, St-Amour G, Zhu T, and Headrick K (1995) Herbicides volatilization measured by the relaxed eddy-accumulation technique using two trapping media. Agric. For. Meteorol. (76): 201–220.

PMEP (2003) http://pmep.cce.cornell.edu./facts-slides-self/core-tutorial/module06/

Polyrakis I (2004) *The Pollution of the Environment from Agrochemicals* (under edition, in Greek).

Rao PSC, Mansell RS, Baldwin LB and Laurent MF (1998) http://pmep.cce.cornell.edu/facts-slides-self/facts/gen-pubre-soil-water.html

Risebrough RW, Hugget RJ, Griffin JJ and Goldsberg ED (1968) Pesticides: transatlantic movements in the northeast trades. Science (159): 1233.

Roumpos ICh, Yannopolitis KN and Broumas Th (2001) The side effects of pesticides on flora, fauna and on useful arthropods. In: *Observations on Environmental Side Effects from the Use of Pesticides in Agriculture.* Current state in Greece. Athens, 2001 (in Greek).

Smith WM Jr. and Ledbetter JO (1971) Hazards from fires involving organophosphorus insecticides. Am. Ind. Hyg. Assoc. J. (32): 468.

Sotherton NW, Dover JW and Rands MRW (1988) The effects of pesticide exclusion strips on faunal populations in Great Britain. Ecol. Bull. (39): 197–199.

Spencer WF and Cliath MM (1990) Movement of pesticides from soil to the atmosphere, In: *Long Range Transport of Pesticides*, Lewis Publ. pp. 1–16.

Tabor EC (1965) Pesticides in urban atmospheres. APCA J. (15): 415.

Tzanakakis ME (1961) *Methods to Control Insects*. General Part. Athens (in Greek).

Van den Berg F, Bor G, Smidt RA, Van den Peppel-Groen AE, Smelt JH, Muller T, and Maurer T (1995) Volatilization of parathion and chlorothalonil after spraying onto a potato crop. SC-DLO, Agricultural Research Department, Winand Staring Centre for Integrated Land, Soil and Water Research, Report 102.

Valavanidis A (2000) *Fundamental Principles of Environnemental Chemistry, Ecotoxicology, and Ecologic Risk Assessment* (in Greek).

Ware GW, Cahill WP, Gerhardt PD and Witt JM (1970) Pesticide drift IV. On-target deposits from aerial application of insecticides. J. Econ. Entomol. (63): 1982.

Waymann B and Rüdel H (1995) Influence of air velocity, application dose, and test area size on the volatilization of lindane. Int. J. Environ. Anal. Chem. (58): 371–378.

Wienhold BJ and Gish TJ (1994) Effect of formulation and tillage practice on volatilization of atrazine and alachlor. J. Environ. Qual. (23): 292–298.

Wolfe HR (1979) Field exposure to airborne pesticides. In: *Air Pollution from Pesticides and Agricultural Processes*. CRC, Boca Raton, FL.

Woodwell GM, Craig PP and Johnson HA (1971) DDT in the biosphere: where does it go? Science (174): 1101.

WSU (1996) http://cru.cahe.wsu.edu/CEPublications/eb1543/eb1543.html

WWF (1999) http://www.neteffect.ca/pesticides/resources/bugs-at-risk.pdf

Yeo D (1959) The problem of distribution; The physics of falling droplets and particles. The drift hazard. 1st Int. Agric. Aviation Conf. 112.

Zouridakis A, Alexandrakis V, and Yamvrias Ch (1995) Study of the factors effecting the action of predators of *Aspidiotus nerii* Bouché (Hom. Diaspididae). Proceedings of the 6th Pan-Hellenic Entomological Conference, Hania, Oct 31–Nov 3, 1995 (in Greek).

Chapter 12
Phytoremediation of Metal-Contaminated Soil for Improving Food Safety

Stefan Shilev, Manuel Benlloch, R. Dios-Palomares, and Enrique D. Sancho

12.1 Introduction

The contamination of the environment is a serious problem which provokes great interest in our society and in the whole scientific community. The input of metals into soils has increased during the last few decades as a consequence of different human activities (storage of industrial and municipal wastes, burning of fuels, mining and wastewater treatments, functioning of non-ferrous-metal-producing smelters, etc.). Nowadays, this type of contamination is one of the most serious concerning the chronic toxic effect which it renders on human health and the environment. As a consequence of all these activities, a huge number of toxic metals and metalloids, such as Cu, Zn, Pb, Cd, Hg and As, among many others, have been accumulated in soils, reaching toxic values. Unfortunately, much contaminated land is still in use for crop production, despite the danger that the metal content poses.

In western Europe more than 1.4 million sites are polluted with heavy metals and metalloids (McGrath et al., 2001). This type of contamination is a big problem, also for agricultural land. Remediation of heavy metal contaminated soils is difficult owing to the lack of possibilities to destroy heavy metal pollutants, while this objective for organic pollutants can be achieved by mineralization in most cases. Present technologies for soil restoration are difficult, too expensive, time-consuming and in many cases create an additional risk for people and produce secondary waste (Wenzel et al., 1999). For that reason it is very important to develop efficient and cost-effective in situ technologies.

To secure the future of the next generations, we need a sustainable source of food. Securing the future supply of high-quality, healthy and safe food and minimizing the impact of contaminants are linked goals. The focus also should be directed to increase essential elements in the edible parts of food crops and significant

Stefan Shilev (✉)
Department of Microbiology and Environmental Biotechnologies,
Agricultural University of Plovdiv, Bulgaria
stefan.shilev@au-plovdin.bg

R. Costa, K. Kristbergsson (eds.), *Predictive Modeling and Risk Assessment,*
DOI: 10.1007/978-1-387-68776-6, © Springer Science+Business Media, LLC 2009

reduction of the uptake of heavy metals and organic pollutants. A better use of tools, such as accumulation and metabolism of pollutants by plants, can be very useful for decreasing the risk of contamination of food via water bodies and soils.

The term "bioremediation" has been defined as a "biological response to environmental abuse" (Hamer, 1994). The purpose of this technology is to utilize natural processes for remediation of contaminated sites. During the last few years, with the development of phytotechnologies, efficient tools and environmental solutions for the cleanup of contaminated sites and waters have been proposed to support the improvement of food safety and sustainable land use management. The use of plants, especially selected for the rehabilitation of polluted lands and water purification, is very important to reach this objective. One bioremediation technology in the development stage is *phytoremediation*. This technology uses, basically, higher plants to accumulate the contaminant in their aboveground parts (phytoextraction) or to immobilize it in the rhizosphere (phytostabilization). On the other hand, cropping plants with reduced capacity to store metal in the edible parts could be very useful for improving food safety.

Generally, soil microorganisms are responsible for more than 90% of soil biological activity and have a direct influence on many physiological aspects of plant growth (regulation of vegetative growth, transport of water and nutrients, plant protection from soil-borne pathogens, etc.). The soil microbiological activity can affect the solubility, accumulation and mobility of metals and metalloids, which is fundamental for phytoremediation. Despite these advantages, little information is available concerning the rhizosphere processes and the participation of microorganisms in phytoremediation of metal- and metalloid-contaminated soils.

The use green plants to clean up the environment is not new. At the end of the eighteenth century, *Thlaspi caerulescens* was the first plant species documented to accumulate high levels of metals in leaves (Baumann, 1885). More recently, Rascio (1977) reported tolerance and high Zn accumulation in shoots of *T. caerulescens*. In the last decade, extensive research has been conducted to investigate the biology of metal phytoextraction (Lasat, 2000). Despite significant success, our understanding of the plant mechanisms that allow metal extraction is still emerging.

Nowadays, the phytoremediation markets occupy an important place in the site-restoration business. Glass (1999) approximates that for phytoremediation of soil contaminated with metals during 1999, the US market was worth about \$4.5 million to \$6 million, while in the EU these values were tenfold smaller.

12.2   Toxic Metals: The Environmental Problem

The heavy metals are inorganic elements with density greater than 5 g cm^{-3}. From 90 elements occurring naturally, 21 are not metals, 16 are light metals and 53 (including metalloids such as As and Se) are heavy metals (Weast, 1984). Some of them take part in numerous biochemical reactions in living organisms (Nies, 1999). This is the reason for the interest regarding their possible toxic effects when toxicity limits are surpassed.

12.2.1 Risk Assessment

Soil remediation is needed to eliminate risk to humans, animals and the environment as a whole. A lot of information exists regarding the toxicity of Cd, Se and Pb in soils (Cai et al., 1990), as well as Zn, Ni and Cu in industrial and mine zones, to sensitive plants (Chaney et al., 1999). The exposition of Pb provokes mutation, cancer, sterility, etc. (Lopez et al., 2002). The presence of high concentrations of As in different zones (Argentina, Bangladesh, Chile, Thailand, Taiwan, etc.) is closely related to an increased frequency of some types of cancer, e.g. skin, bladder, liver, kidney and lung (Morales et al., 2000; Steinmaus et al., 2000).

12.2.2 Bioavailability or Toxicity

The toxicity of heavy metals is strongly related to their bioavailability (bioaccessibility) – the proportion of a chemical presented in accessible form to organisms. The bioavailable fraction is very important from the ecotoxicological and health point of view and depends upon the chemical soil properties of pollutant, ages, climate and biological behaviour. Ageing usually leads to a decrease in bioavailability; nevertheless, root exudates, mycorrhizal fungi and rhizospheric bacteria play an important role in the ability of pollutants to enter plants and enhance toxicity in all living organisms. Conversely, the reduction of Hg to the ionic or elemental form will render the metal less bioavailable and less toxic. The reduction of arsenite (As^{3+}) to arsenate (As^{5+}) leads to a decrease of toxicity. However, there is not enough information on how to assess the bioavailability of many metallic elements. It can be difficult to quantify the bioavailability, because there are many factors involved, such as pH, soil moisture, organic matter content, soil type (e.g. carbonate-rich soils) and other compounds, that affect it.

Bioavailability of pollutants and their uptake by crops are essential parameters for establishing the risk and enhancing food safety, although this is a controversial area in both regulation and remediation. The assumption that 100% of the metal bioavailability means 100% bioavailability to plants often overestimates the impact of the metals. The solubility plays an important role in the bioavailability of metals to plants. Many inorganics are presented in insoluble forms and require the addition of chemical amendments (chelates) to make them more soluble. The use of chelates results in a huge increase of the ability of plants to accumulate metals, and after that the plant biomass can be harvested for recovery and disposal.

Variance in toxicity level depends upon the nature of the metal. Some heavy metals (Co, Cu, Fe, Mn, Mo, Zn) are essential for biological processes since they are included as structural and catalytic components of proteins and enzymes, while others (As, Cd, Hg, Pb) do not participate in cellular metabolic processes. The metals from both groups, at high concentrations, could be very toxic to any living organism, when displaced metallic cofactors react with essential groups of proteins and coenzymes or promote the formation of reactive species of oxygen. In this sense, we have to include some non-metal elements (As, Se) to complete the list of the most "famous" inorganic contaminants (Weast, 1984).

12.3   Overview of Phytotechnologies

Phytotechnologies use green plants to remediate contaminated sites. The remediation can be applied in situ (on the site where contamination had been produced) or ex situ (excavate the contaminated media from location). Generally, the contaminant can be of organic (petroleum hydrocarbons, gas condensates, crude oil, chlorinated compounds, pesticides, explosives, etc.) or inorganic (inorganic salts, heavy metals, metalloids and radioactive materials) nature. The media affected include soils, sediments, groundwater and surface waters.

12.3.1   Fundamentals

Plant physiological function is well known. Plants grow by sending their roots into the soil (commonly). They utilize carbon dioxide to photosynthesize carbon biomass, produce energy and release oxygen to the environment, take up and transpire water from the subsurface, absorb dissolved inorganics through the root system and exude photosynthetic products into the rhizosphere (Taiz and Zeiger, 1991).

The essential inorganic plant nutrients (N, P, K, Ca, Mg, S, Fe, Cl, Zn, Mn, Cu, B, Mo) are taken up by the root system as dissolved components from the soil solution. These elements are fundamental for plant growth and metabolic processes. When they are in the root system, the dissolved elements can be transported through the xylem. Apart from the essential nutrients, other non-essential elements (salts, Cd, Pb, As, etc.) can also be taken up.

The compounds exuded by plants into the rhizosphere are amino acids, enzymes, proteins, organic acids, carbohydrates, etc. The exuded carbon could be around 20% of the total products synthesized by the plant (Taiz and Zeiger, 1991). Soil microorganisms (bacteria and fungi) are in the immediate vicinity of the root hairs, because surrounding soil is enriched with the abovementioned nutrients. This soil is known as the rhizosphere and it extends several millimetres from the root hair. The symbiotic relationships within the rhizosphere are very complicated and are not well known yet. Selected strains of *Pseudomonas* sp. and *Bacillus* sp. have been shown, in a hydroponic system, to increase the total amount of Cd accumulated in Indian mustard (Salt et al., 1995) or of As in sunflower (author's unpublished results), although the results of hydroponic trials are not necessarily representative for soil conditions.

Phytotechnologies are developed to remediate contaminated soils with organic or inorganic pollutants, to improve food safety and to increase food quality. Generally, there are six phytoremediation processes (the common term for utilization of phytotechnologies on contaminated lands): phytostabilization, phytoimmobilization, phytovolatilization, phytodegradation, rhizofiltration and phytoextraction. The first two processes are containment processes, while the rest are removal processes.

12.3.2 Phytostabilization

Phytostabilization is a containment technology that uses plants tolerant to pollutants with the purpose of stabilizing mechanically the polluted soils by plant roots, to prevent erosion and leaching. This process requires adaptation of plant species to the contaminant concentrations in the soil and good developed roots.

There are three phytostabilization mechanisms that determine the fate of pollutants:

1. Phytostabilization in the root zone – the contaminant is immobilized by the plant root exudates (amino acids, sugars, enzymes) released into the surrounding soil and decreasing the bioavailable soil fraction.
2. Phytostabilization on the root membrane – binding the pollutant to root surface through root-associated proteins, preventing plant uptake.
3. Phytostabilization in root cells – includes root cell sequestration of the contaminant in the vacuole. Probably this process is related to chelation by proteins.

12.3.3 Phytoimmobilization

This technology uses plants to decrease bioavailability and mobility of metals and metalloids in the root zone, preventing plant uptake. Two plant-based mechanisms have been described – root adsorption/absorption and plant-assisted formation of insoluble compounds (Wenzel et al., 1999). Alteration of soil factors also could be an important part of the immobilization technology. On the other hand, rhizosphere microorganisms may participate in the immobilization through fixation (adsorption and uptake) or synthesis of less mobile compounds. Phytoimmobilization is a very important process because reduction of pollutant plant uptake to the shoots avoids transfer into the food chain. In this sense, screening of food cultivars is necessary to select species which accumulate metals at the lowest possible amounts.

12.3.4 Phytovolatilization

Phytovolatilization is a promising process for elimination of some metallic (as well as organic) contaminants absorbed by the roots and volatilized into the atmosphere. Nevertheless, the fact that it is possible to apply it just for a few elements (Hg, Se and possibly As) (De Souza et al., 1999, 2002) demonstrates the basic role of accumulation processes.

12.3.5 Phytodegradation

Phytodegradation is an alternative to conventional microbial biodegradation or plant use to stimulate microbial decontamination. The process is applicable just for organic pollutants, while metals and metalloids cannot be destroyed. Normally, the plants participate in the mobilization of the contaminant into the rhizosphere and this helps microorganisms in the biodegradation process.

12.3.6 Rhizofiltration

Rhizofiltration refers to the use of plants to absorb contaminants from solutions, accumulating them in the plant roots. This process is not directly related to soil bioremediation. It is based on the ability of plants to absorb into the roots organic or inorganic contaminants.

12.3.7 Phytoextraction

Plants have evolved the ability to acquire nutrient ions and organic compounds at low concentrations from a given environment and accumulate them in the roots and shoots. This technology is known as phytoextraction. Nowadays, the accumulation of mineral elements by plants has attracted great interest, owing to the possibility of utilization as a decontamination technology for toxic metals and metalloids. Phytoextraction is one of the most efficient, secure and low-cost alternatives for remediation of soils contaminated with common metals (Ag, Al, Cd, Cr, Pb, Hg, Sn, Zn, Cu, etc.) and metalloids (As, Se).

There exist two general categories of phytoextraction, based on utilization of different types of plants: continuous and induced phytoextraction.

12.3.7.1 Continuous Phytoextraction

Continuous phytoextraction is based on the utilization of hyperaccumulating plants which have physiological possibilities to absorb and accumulate metals at very high concentrations throughout their biological cycle. Those plants are defined as hyperaccumulators, which can store in their aerial parts metallic contents above 0.1% dry weight ($1,000\ \mu g\ g^{-1}$ dry weight) for almost all elements, except for Zn and Mn (1%, 10,000 $\mu g\ g^{-1}$), Cd (0.01%, 100 $\mu g\ g^{-1}$) and Au (0.0001%, 1 $\mu g\ g^{-1}$) (Brooks et al., 1998).

Although they are not very popular, hyperaccumulating plants have been known for a long time. After their recent rediscovery (Brooks et al., 1977), the number of known species has increased considerably. Currently more than 400 species of hyperaccumulating plants have been described (Baker et al., 2000; Brooks, 1998) (Table 12.1).

Table 12.1 Heavy metal hyperaccumulating plants (Brooks, 1998)

Element	Species	Families
Cadmium	1	Brassicaceae
Cobalt	26	Lamiaceae, Scrophulariaceae
Copper	24	Cyperaceae, Lamiaceae, Poaceae, Scrophulariaceae
Manganese	11	Aprocynaceae, Cunoniaceae, Proteaceae
Nickel	290	Brassicaceae, Cunoniaceae, Euphorbiaceae, Flacourtiaceae, Violaceae
Selenium	19	Fabaceae
Thallium	1	Brassicaceae
Zinc	16	Brassicaceae, Violoceae

The continuous phytoextraction has several limitations that decreases its practical utility (Salt et al., 1998; McGrath et al., 1995), and these are related to low biomass production and growing rate. Both factors determine an insufficient level of remediation, mainly owing to the lack of specific hyperaccumulators for important metallic pollutants such as As, Cd and Pb (Cong and Ma, 2005).

Although the first limitation (low biomass production) could be alleviated through agronomical practices to optimize the cropping of this type of plants, the necessity to expand and improve the known hyperaccumulators seems clear, as well as to increase the available number of these plants supplied with huge biomass. Furthermore, we have to consider the implication in this technology of the rhizosphere microbes, whose role is not very clear yet.

12.3.7.2 Induced Phytoextraction

During the last few years many investigations have been directed to the utility of agricultural plants tolerant to metal ions and possessing huge biomass, although they do not accumulate high concentrations of metals in the shoots. One of these plants is the Indian mustard (*Brassica juncea*). This is a cultivated species with high biomass which had been used for experiments of phytoremediation of soil contaminated with Se, and phytoextraction of Cd, Cu, Ni, Pb and Zn (Blaylock et al., 1997; Salt et al., 1995, 1998). Attention has now turned to numerous crop species such as rice, oat, barley, sunflower, pea, maize, beetroot, tobacco, tomato and wheat (Simon, 1998; Shilev et al., 2000).

The phytoextraction of crop plants with high biomass is inconvenient in some respects, such as low absorption rate and/or metal translocation to the shoots or insufficient tolerance to the metals. A lot of heavy metals and metalloids have low availability for plants and, in addition, the metal is retained in the roots.

The abovementioned restrictions are especially applicable to Pb. In many cases the usage of chelators is a successful tool for mobilization of this heavy metal. The application of synthetic chelates to the rhizosphere of Indian mustard has been a key factor to achieve Pb accumulation of up to 1.6% in the shoots, concurring with increments of absorption and translocation as well as increase of concentration in soil solution (Blaylock et al., 1997). In most cases the relative increment of accumulated Pb achieved by chelating agents was much higher in the stems than in the roots. This indicates that there exists stimulation, not just of the absorption of the metal, but also of its translocation from the root to the shoot. These effects of the chelating agents also extend to other metals (Cd, Cu, Ni and Zn), although the interactions in the plant–rhizosphere–metal system are very complicated and each case has particular characteristics depending on many physical and chemical factors (Luo et al., 2005).

The utilization of chelates could permit the establishment of the different growth phases from metal phytoextraction, resulting in the technique known as induced phytoextraction (Salt et al., 1998; Vassilev, 2002).

Although big biomass plants do not show high levels of absorption or translocation capacity of heavy metals, nor are they very tolerant to them, this technology allows the cultivation of these plants on contaminated sites until a determined phase of growth is reached; afterwards, the chelating agent promoting the heavy metal absorption can be applied. Finally, this is followed by the harvesting of plant biomass for extraction of accumulated metals.

12.3.7.3 Determinative Factors for Metal Phytoextraction

Metal Bioavailability

The success of remediation of metal-contaminated soils based on phytoextraction or phytoimmobilization depends mainly on the bioavailability of contaminants (Salt et al., 1995). There are different states of the metal in the soil: (1) simple metal ions and soluble complexes present in soil solution; (2) metal ions bound to inorganic soil components (e.g. places of ionic exchange); (3) metals bound to soil organic matter; (4) precipitated or insoluble salts (e.g. carbonates, oxides, hydroxides, etc.); (5) as part of silicate structure. The last fraction is the native soil content, while the rest correspond to anthropogenic activities. Fractions available to plants in normal conditions for the phytoremediation (extraction) processes are fractions 1 and 2.

In addition, different rhizosphere microorganisms have an important role in the phytoassimilation of metal micronutrients and could increment the metal absorption capacity of the plants. Plant growth promoting bacteria are very useful in the processes of metal phytoextraction as well. They improve plant growth on contaminated soils and participate in transformation of metals (e.g. volatilization or reduction) to more available forms for plants. In this sense, De Souza et al. (1999) demonstrated that rhizobacteria had contributed up to 35% of the volatilized and up to 70% of the accumulated metalloid by Indian mustard's (*B. juncea*) tissues.

Tolerance to Heavy Metals and Metalloids

Plants. The heavy metals and metalloids are not essential and, at high concentrations, can operate as toxins. In this sense, plants contain different mechanisms to stabilize or sequester them and to prevent translocation into sensitive tissues: exclusion; complex formation; compartmentalization (vacuole); biotransformation and cell wall reparation (Salt et al., 1998).

One of the mechanisms leading to decrease of metal toxicity in plants is complexation (binding free ions from solution in less toxic complexes). In plants this is possible through production of peptides – metallothioneins (MT) (Robinson et al., 1993) or phytochelatins (PC) (Rauser, 1995). They are characterized by containing a high quantity of the amino acid cysteine and are also thiol-rich. From a biochemical point of view, the difference between them consists in their formation; while the MTs are genetic products and are produced from the proteins, the PCs are obtained during the metabolism of glutathione (–GluCysGly, GSH). The MTs were firstly described in mammals and after that in other animals and fungi. In plants just one protein identified as a MT (Ec in wheat germ) has been obtained, while in different species genes encoding proteins of MT type have been found. The PC structure was identified as $(-GluCys)_n Gly$ ($n = 2$–11, depending on the organism) (Robinson et al., 1993). In plants the detoxification of heavy metals is done by PCs. In Indian mustard, the increment of levels of GSH and PCs is accomplished through overexpression of enzymes for biosynthesis of GSH (glutathion synthetase and g-glutamylcysteine synthetase) (Zhu et al., 1999).

The compartmentalization of some metals or their complexes with appropriate ligands is a normal mechanism of tolerance occurring basically in the vacuole. The immobilization on cell wall and in leaf trichomes is another mechanism of tolerance which could explain tolerance to some metals (Salt et al., 1998).

The biotransformation is an additional mechanism of tolerance. It is known that some elements (e.g. Se; De Souza et al., 2002) can be volatilized by plants and others (e.g. Hg) by microorganisms (Moreno et al., 2005).

Cell reparation is another mechanism for detoxification in some oxidative stress conditions. It is known that some metals cause oxidative stress in plants owing to a decrease of the concentration of intracellular GSH (production of PCs) or production of reactive oxygen forms. In this case the activation of some mechanisms for metal protection is possible, related to chemical (GSH, ascorbate or phenolic compounds) or enzymatic antioxidants (catalase, peroxidase, glutathione reductase, ascorbate peroxidase).

Microorganisms. The bacterial community of soil is very diverse, and there are more than 13,000 bacterial species, with an indeterminate number of fungi, algae and protozoa (Torsvik et al., 1990). Generally, the changes produced by the heavy metals in the soil microbial community and its diversity are not well known, although it has been described that the microorganisms which survive in long-term contaminated sites represents 10–100% of the community structure and this demonstrates big diversity, dominated by gram-negative bacteria (Kunito et al., 1998, 2001; Giller et al., 1998; Francis, 1999), basically represented by *Pseudomonas* and *Flavobacterium*.

The reasons for adaptation of the microbiota to the metal content of the soil could vary a lot depending on the decrease of bioavailability of metals during the time or changes produced in the microbial community owing to an increase in tolerance to those metals (Holtan-Hartwig et al., 2002). Bacterial tolerance to the heavy metals normally is due to spontaneous mutations as well as transference of resistance genes through plasmids (Osborn et al., 1997). This is an original characteristic of some microbial strains or an acquired ability of populations caused by growth in specific conditions. In some natural soils the plasmid transfer and the plasmid stability is a distinguishing feature.

Moreover, microorganisms are responsible for more than 80% of soil activity and for the transformation of at least one third of the elements of the periodic table. These transformations are related to valence (redox processes) and state (solid, liquid, gaseous) changes, and play an important role in many biochemical processes. Under appropriate conditions, the solubilization or immobilization of metals is directed by one or more than one of following mechanisms: oxidation–reduction; pH changes (affects ionic state and solubilization of the metals); dissolution produced by organic acids and other metabolites; volatilization; immobilization through complex formation; bioaccumulation or biopolymer formation as well as biotransformation of the complexes between metals and organic compounds (Nies, 1999; Sharples et al., 2000).

The direct action of metal solubilization is due to the production of mineral–organic acids by autotrophic bacteria as well as to the production of organic acids by organic metabolism by heterotrophic bacteria (Francis, 1999). These bacteria and fungi liberate metals from different minerals. In some cases, the combined effect of pH reduction and complex formation is an important tool for metal solubilization. Some acids of microbiological nature (dicarboxylic; salicylic, etc.) function as effective chelating agents of heavy metals by increasing their mobility in soil.

12.3.7.4 Agronomic Practices To Enhance Metal Phytoextraction

Although phytoextraction has some inconveniences related to the low biomass of hyperaccumulating plants or insufficient plant-tissue concentration of metals and metalloids, it is possible to improve it through optimization of agronomical practices for those plants. Other criteria are also important for ecosystem protection. As a general rule, native plants are preferred to exotic ones, which could represent some danger for the ecosystem's harmony.

The rate of metal removal depends upon the total extracted metal per unit of soil (e.g. hectare). One of the most problematic subjects is the type of remediative species: metal hyperaccumulators versus common non-hyperaccumulating crop species. The answer depends on a lot of parameters: type of metal which will be extracted; tolerance of plant species to those metals; level of shoot accumulation; shoot biomass formation; possibilities for several cropping per year or mechanized culturing and harvesting; speed of growth, etc.

There are many factors involved in the optimization of phytoextraction: plant selection; soil fertilization and conditioning (N, S, P, K, etc.); enhancement of metal availability by synthetic chelators [EDTA, N-(hydroxyethyl)ethylenediaminetriacetic acid, diethylenetriaminepentaacetic acid]; sowing; crop rotation; crop maintenance (pest control, water supply); cost- and time-effectiveness (Lasat, 2000). Each one is important for improving the results by in situ phytoextraction and has to be supervised.

12.4   Phytoextraction of Arsenic (Case Study)

12.4.1   Introduction

The investigation we present here was part of a research project on bioremediation of metal-contaminated soils carried out by a multidisciplinary group from the Agronomy School (University of Cordoba, Spain) (Shilev et al., 2003). The contamination originated from an environmental disaster of great dimensions which took place in April 1998 near Aznalcóllar, Seville province, southern Spain. At that time, the failure of the tailings pond dam at the Aznalcóllar pyrite mine caused the release of 4.5×10^6 m^3 of toxic spill containing distinct metals and metalloids (As, Cd, Cu, Fe, Pb and Zn, basically). Approximately 4,300 ha along 40 km of the Agrio and Guadiamar valleys was affected. The contamination extended to the Doñana National Park, which is in the vicinity, affecting soils, waters, animals and plants (Serrano, 1999). After the first steps related to retention of the spill and its excavation had been made, efforts were directed to the remediation of the affected sites.

Understanding the relations between all parts involved in the plant–rhizosphere–soil system is very important to enhance the phytoextraction of such pollutants as the metalloids, particularly As. A regression mathematical model to describe the dependence between bacterial inoculation, As in soil and the amount of metalloid extracted was developed.

12.4.2   Materials and Methods

The study was carried out in a greenhouse over 35 days. Two-litre pots filled with uncontaminated soil and plant material of sunflower (*Helianthus annuus* L. 'Sun-Gro 393') were used. One week after seed germination, two arsenic concentrations ($Na_2HAsO_4 \cdot 7H_2O$), 5 and 20 mg As L^{-1}, were prepared, and a control treatment without As (0 mg As L^{-1}), was used. Each the treatments was repeated with addition of bacteria *Pseudomonas fluorescens* (which had been isolated previously), tolerant to metals, in a concentration of 10^6 cfu mL^{-1} of soil (cfu is colony forming units). In the treatments without bacteria exhausted medium was added. Each treatment had four replications. Water and nutrient solution were regularly added. Growth parameters (fresh and dry weight, height, leaf surface and volume of xylem fluxes) were measured. The analysis of total As accumulated in the plant tissues as well as in xylem fluxes was measured by atomic absorption spectroscopy.

Until now several mathematical models on root uptake of cations, metal movement in soil, surface flows, etc. have been made (Darrah and Staunton, 2000), although little information is available on the prediction of the accumulation rates and on the relations between the heavy metal content of soil and shoot concentration. In this investigation, a regression mathematical model was used with the purpose of predicting the As accumulation in shoots of sunflower in phytoextraction and bioaugmentation assays. In this model a Student's t distribution with degrees of freedom was used.

12.4.3   Results and Discussion

In this experiment we studied the effect of inoculation of bacteria, *P. fluorescens* tolerant to heavy metals and metalloids, on the growth and accumulation of As in sunflower plants, grown in non-contaminated soil with addition of this metalloid. On the basis of the results, a regression mathematical model was obtained for studying the influence of two independent variables (As concentration in soil and presence of *P. fluorescens*) on the dependent variables (As accumulated in stems, leaves or roots and the volume of xylem fluxes).

Effect of As in soil and P. fluorescens on As concentration in the stems.

The As concentrations in stems in both treatments, with As and without bacteria (Fig. 12.1), were similar despite differences in soil As values. On the other hand, the inoculation of the rhizobacterium *P. fluorescens* in the treatment with a higher dose of As had a positive effect on the accumulation in stems, where the increment was more than 2 times higher than in the treatment without bacteria.

According to the regression model mentioned earlier, we obtained the follow dependency:

$$\text{As}(\text{stem}, \mu g\, g^{-1}) = -0.074 + 0.193 \times \text{arsenic} + 1.382 \times \text{bacteria}$$
$$(-0.558)\ (6.627)\ (2.795).$$

The numbers in parentheses are t statistics.

Both variables were significantly distinct from zero, with significance of 1%, where $R^2 = 0.684$ and F from the analysis of variance (ANOVA) demonstration model was significant to a level below 1‰. The presence of bacteria in soil supports the increase of As concentration in the stems and contributes to an increment of 1.382 µg As g^{-1} dry weight. In respect to the presence of As in soil, it is obvious that on increasing the soil concentration to 1 mg L^{-1}, the stem concentration rises to 0.193 µg As g^{-1} dry weight.

Effect of As in soil and P. fluorescens on arsenic concentration in the leaves.

The As concentration in leaves was between 4 and 5 times higher than in the stems (Fig. 12.2). In both treatments with addition of As, without bacteria, the concentrations of As were very similar, while in the treatments with bacteria the accumulated concentrations were incremented by 60 and 100%, respectively. In the mathematical model

$$As\,(leaves, \mu g\,g^{-1}) = 1.099 + 0.629 \times arsenic + 5.141 \times bacteria$$
$$(0.685)\ (5.732)\ (2.758).$$

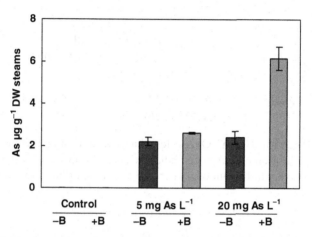

Fig. 12.1 Effects of arsenic in soil and *Pseudomonas fluorescens*, tolerant to metals and metalloids, on the arsenic concentration in the stems. *DW* dry weight, *–B* without *P. fluorescens*, *+B* with *P. fluorescens*

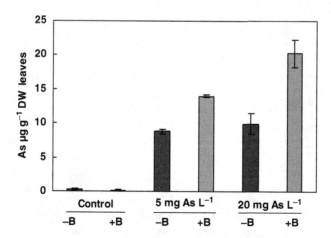

Fig. 12.2 Effects of arsenic in soil and *P. fluorescens* on the arsenic concentration in the leaves

Both explanatory variables were significantly distinct from zero, with a significance of 1%, where $R^2 = 0.626$ and F from the ANOVA demonstration model was significant to a level below 1‰. This means that the inoculation of the bacterial population supports the increment of As to 5.141 µg g^{-1} dry weight of leaves and 1 µg As L^{-1} of soil induces an increment of the leaf concentration of 0.629 µg As g^{-1} dry weight.

Effect of As in soil and P. fluorescens on As concentration in the roots.

In all treatments we observed that on increasing the soil As concentration the As concentration in the roots increased too. On the other hand, the inoculation of bacteria provoked a major accumulation in the roots (4 times when the soil concentration was 5 mg As L^{-1}, and more than 8 times in the treatment with 20 mg As L^{-1}) (Fig. 12.3). These values indicate a strong toxicity effect on the plants, which resulted in significant reduction of plant growth (results not shown).

In this sense the regression model is as follows:

$$As(root, \mu g\,g^{-1}) = -46.830 + 7.002 \times arsenic + 80.163 \times bacteria$$
$$(-2.412) \quad (5.276) \quad (3.553),$$

where $R^2 = 0.626$ and F of ANOVA is less than 1‰, indicating that the presence of tolerant bacteria promotes As accumulation in the roots was 80.163 µg g^{-1} dry weight, while every milligram of As per litre of soil contributes to 7.002 µg As g^{-1} dry weight

Effect of As in soil and P. fluorescens on the xylem fluxes of the plant.

The presence of As in the soil in both concentrations did not affect the xylem fluxes of plant, while the addition of *P. fluorescens* promoted the increment of the volume of xylem fluxes in the treatments with bacteria (with and without As) (Fig. 12.4) although the increment in the treatment with 20 µg As L^{-1} was very small. According to the mathematical regression model

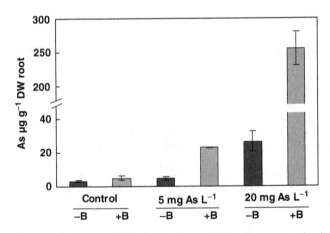

Fig. 12.3 Effects of arsenic in soil and *P. fluorescens* on the arsenic concentration in the roots

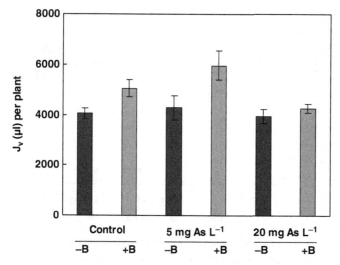

Fig. 12.4 Effects of arsenic in soil and *P. fluorescens* on the xylem fluxes (*JV*), expressed per plant

$$\text{Fluxes (µl)} = 4{,}092.5 + 995.5 \times \text{bacteria}$$
$$(17.902)\ (3.079)$$

Only the presence of rhizobacteria gave values significantly distinct from zero, with a significance of 5‰, where $R^2 = 0.269$ and F of the ANOVA demonstration model was significant to a level below 5–1,000. This indicates that the presence of metalloid did not affect the xylem fluxes, whereas the presence of bacteria contributes to an increment of 995.5 mL per plant.

Effects of As in soil and P. fluorescens on total As accumulation in the roots.

The total As accumulated in the root depends on the concentration in the tissue and on the total weight of the roots. In this sense, although the root weight was reduced in the treatment with 20 µg As L^{-1} with *P. fluorescens*, the total accumulated quantity of As was much higher than in the rest of the treatments (Fig. 12.5).

$$\text{As accumulated (root, µg)} = -1.778 + 0.273 \times \text{arsenic} + 3.208 \times \text{bacteria}$$
$$(-1.842)\ (4.141)\ (2.859)$$

In the mathematical model of total As accumulated in the root

The addition of bacteria increases the As accumulation in the roots and contributes to an increase of 3.208 µg As per root. In respect to the presence of As in soil, 1 µg As L^{-1} contributes to the accumulation with 0.273 µg of As.

Although significant reduction of plant growth was observed in the treatment with the highest As concentration, the volume of xylem fluxes was not reduced. The fact that the accumulation of As in the roots was so high could explain the inhibition of shoot growth mentioned earlier. In this sense, the presence of tolerant

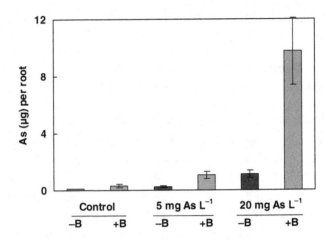

Fig. 12.5 Effect of arsenic in soil and *P. fluorescens* on the total arsenic accumulated into the roots

bacteria ameliorates the negative effect of As toxicity, so the total amount of As extracted into the roots was greater in this treatment.

Finally, the presence of the rhizobacterium *P. fluorescens* clearly promoted the As accumulation in stems, leaves and roots. The fact that the plant biomass was enhanced following the inoculation of rhizobacteria tolerant to metals (*P. fluorescens*) could be considered as a plant growth promoting rhizobacteria effect. These observations mean that the bacteria can be used as a suitable tool in the phytoextraction protocols.

References

Baker, A.J.M., S.P. McGrath, R.D. Reeves, J.A.C. Smith. (2000) Metal hyperaccumulator plants: a review of the ecology and physiology of a biological resource for phytoremediation of metal-polluted soils. In: N. Terry and G. Bañuelos (eds.), Phytoremediation of contaminated soils and waters, pp. 85–107. Lewis Publishers, Boca Raton, FL.

Baumann, A. (1885) Das Verhalten von Zinksatzen gegen Pflanzen und im Boden. *Landwirtsch. Vers.-Statn*, 31: 1–53.

Blaylock, M.J., D.E. Salt, S. Dushenkov, O. Zakharova, C. Gussman, Y. Kapulnik, B.D., Ensley, I.Raskin (1997) Enhanced accumulation of Pb by Indian mustard by soil-applied chelating agents. *Environ. Sci. Technol.*, 31: 860–886.

Brooks, R.R. (1998) General introduction.In: R.R. Brooks (ed.), Plants that hyperaccumulate heavy metals, pp. 1–15. CAB International, Oxon, UK.

Brooks, R.R., J. Lee, R.D. Reeves, T.Jaffre(1977) Detection of nickelferrous rocks by analysis of herbarium specimens of indicator plant. *J. Geochem. Explor.*, 7: 49–77.

Brooks, R.R., M.F. Chambers, L.J. Nicks, B.H. Robinson (1998) Phytomining. *Trends Plant Sci.*, 9: 359–362.

Cai, S., Y. Lin, H. Zhineng, Z. Xianzu, Y. Zhalou, X. Huidong, L. Yuanrong, J. Rongdi, Z. Wenhau, Z. Fangyuan (1990) Cadmium exposure and health effects among residents in an irrigation area with ore dressing wastewater. *Sci. Total Environ.*, 90: 67–73.

Chaney, R.L., Y.M. Li, J.S. Angle, A.J.M. Baker, R.D. Reeves, S.L. Brown,F.A. Homer, M. Malik, M. Chin (1999) Improving metal hyperaccumulators wild plants to develop commercial phytoextraction systems: approaches and progress. In: N. Terry and G.S. Bañuelos (eds.), Phytoremediation of contaminated soil and water. CRC Press, Boca Raton, FL.

Cong, T., L.Q. Ma (2005) Effects of arsenic on concentration and distribution of nutrients in the fronds of the arsenic hyperaccumualtor *Pteris vittata* L. *Environ. Pollut.*, 135: 333–340.

Darrah, P.R., S. Staunton (2000) A mathematical model of root uptake of cations incorporating root turnover, distribution within the plants, and recycling of absorbed species. *Eur. J. Soil Sci.*, 51: 643–653.

De Souza, M.P., D. Chu, M. Zhao, A.M. Zayed, S.E. Ruzin, D. Schichnes, N.Terry (1999) Rhizosphere bacteria enhance selenium accumulation and volatilisation by Indian mustard. *Plant Physiol.*, 119: 565–573.

De Souza, M.P., I.J. Pickering, M.Walla, N. Terry (2002) Selenium assimilation and volatilization from selenocyanate-treated Indian mustard and muskgrass. *Plant Physiol.*, 128: 625–633.

Francis, A.J. (1999) Bioremediation of radionuclides and toxic metal contaminated soils and waters. In: D.C. Adriano (ed.), Bioremediation of metal contaminated soils. American Society of Agronomy, Madison, WI.

Giller, K.E., E. Witter, S.P. McGrath (1998) Toxicity of heavy metals to microorganisms and microbial processes in agricultural soils: a review. *Soil Biol. Biochem.*, 30: 1389–1414.

Glass, D.J. (1999) U.S. and International Markets for Phytoremediation. 1999–2000. D. Glass Assoc. Inc., Needham, MA.

Hamer, G. (1994) Bioremediation: a response to gross environmental abuse. *Trends Biotechnol.*, 11, 317–319.

Holtan-Hartwig, L., P. Dörsch, L.R. Bakken (2002) Low temperature control of soil denitrifying communities: kinetics of N2O production and reduction. *Soil Biol. Biochem.*, 34: 1797–1806.

Kunito, T., H. Oyaizu, S. Matsumoto (1998) Ecology of soil heavy-metal resistant bacteria and perspective of bioremediation of heavy metal-contaminated soils. *Rec. Res. Dev. Agric. Biol. Chem.*, 2: 185–206.

Kunito, T., K. Saeki, K. Nagaoka, H. Oyaizu, S. Matsumoto (2001) Characterization of copper-resistant bacterial community in rhizosphere of highly copper-contaminated soil. *Eur. J. Soil Biol.*, 37: 95–102.

Lasat, M.M. (2000) Phytoextraction of metals from contaminated soil: a review of plant/soil/metal interaction and assessment of pertinent agronomic issues. *J. Hazard. Subst. Res.*, 2: 1–25.

Lopez Alonso, M., J.L. Benedito, M. Miranda, C. Castillo, J. Hernández, R. F. Shore (2002) Interactions between toxic and essential trace metals in cattle from a region with low levels of pollution. *Arch. Environ. Contam. Toxicol.*, 42: 165–172

Luo, Ch., Z. Shen, X. Li (2005) Enhanced phytoextraction of Cu, Pb, Zn and Cd with EDTA and EDDS. *Chemosphere*, 59: 1–11.

McGrath, S.P., A.M. Chaudri, K.E. Giller (1995) Long-term effects of land application of sewage sludge: soil, microorganisms and plants. *J. Ind. Microbiol.*, 14: 94–104.

McGrath, S.P., F.-J. Zhao, E. Lombi (2001) Plant and rhizosphere processes involved in phytoremediation of metal-contaminated soils. *Plant Soil*, 232: 207–214.

Morales, K.H., L. Ryan, T.L.Kuo, M.M. Wu, C.J. Chen (2000) Risk of internal cancers from arsenic drinking water. *Environ. Health Perspect.*, 108: 655–661.

Moreno, F.N., C.W.N. Anderson, R.B. Stewart, B.H. Robinson (2005) Mercury volatilisation and phytoextraction from base-metal mine tailings. *Environ. Pollut.*, 136: 341–352.

Nies, D.H. (1999) Microbial heavy-metal resistance. *Appl. Microbiol. Biotechnol., 51*, 6: 730–750.

Osborn, A.M., K.D. Bruce, P. Strike, D.A. Ritchie (1997) Distribution, diversity and evolution of the bacterial mercury resistance (mer) operon. *FEMS Microbiol. Rev.*, 19: 239–262.

Rascio, W. (1977) Metal accumulation by some plants growing on Zn mine deposits. *OIKOS*, 29: 250–253.

Rauser, W.E. (1995) Phytochelatins and related peptides: structure, biosynthesis, and function. *Plant Physiol.*, 109: 1141–1149.

Robinson, N.J., A.M. Tommey, C. Kuske, P.J. Jackson (1993) Plant metallothioneins. *J. Biochem.*, 295: 1–10.

Salt, D.E., M. Blaylock, N.P.B.A. Kumar, V. Dushenkov, B.D. Ensley, I. Chet, I. Raskin (1995) Phytoremediation: a novel strategy for the removal of toxic metals from the environment using plants. *Biotechnology*, 13: 468–474.

Salt, D.E., R.D. Smith, I. Raskin (1998) Phytoremediation. Annu. Rev. Plant Physiol. *Plant Mol. Biol.*, 49: 643–668.

Serrano, J. (1999) Balance de las actuaciones realizadas en relación con el minero de Aznalcóllar. Seminario Internacional sobre Corredores Ecológicos y Restauración de Ríos y Riberas. Resúmenes Ponencias, pp. 13–14. Junta de Andalucía. Consejería de Medio Ambiente.

Sharples, J.M., A.A. Meharg, S.M. Chambers, J.W.G. Cairney (2000) Symbiotic solution to arsenic contamination. *Nature*, 404: 951–952.

Shilev, S., M. Benlloch, E.D. Sancho (2000) Effects of rhizospheric bacteria on heavy metals extraction by sunflower (*Helianthus annuus*). Plant Physiol. *Biochem.*, 38 (Suppl): S19–S81. 12th FESPP Congress, Budapest, Hungary, pp. 15–18.

Shilev, S., M. Benlloch, E. Sancho (2003) Utilization of rhizobacteria *Pseudomonas fluorescens* in phytoremediation strategies. In: T. Vanek and J.-P. Schwitzguebel (eds.), Phytoremediation Inventory COST Action 837 view, p. 39. VOCHB AVČR, Czech Republic.

Simon, L. (1998) Cadmium accumulation and distribution in sunflower plants. *J. Plant Nutr.*, 21: 341–352

Steinmaus, C., L. Moore, C. Hopenhayn-Rich, M.L.Biggs, A.H. Smith (2000) Arsenic in drinking water and bladder cancer. *Cancer Invest.*, 18: 174–182.

Taiz, L., E. Zeiger (1991) Plant physiology. The Benjamin/Cummings Publishing Company, Inc., Redwood City, CA.

Torsvik, V., J. Goksoyr, F.L. Daae (1990) High diversity in DNA of soil bacteria. Appl. *Environ. Microbiol.*, 56: 782–787.

Vassilev, A. (2002) Metal phytoextraction: state of the art and perspectives. *Bulg. J. Agric. Sci.*, 8: 125–140.

Weast, R.C. (1984) CRC handbook of chemistry and physics, 64 edition. CRC Press Inc., Boca Raton, FL.

Wenzel, W.W., D.C. Adriano, D. Salt, R. Smith (1999) Phytoremediation: a plant-microbe-based remediation system. In: D.C. Adriano, J.-M. Bollang, W.T. Frankenberger Jr, R.C. Slims (eds.), Bioremediation of contaminated soils, pp. 457–508. American Society of Agronomy, Madison, WI.

Zhu, Y.L., E.A.H. Pilot-Smith, A.S. Tarun, S.U. Weber, L. Jouanin, N. Terry (1999) Cadmium tolerance and accumulation in Indian mustard is enhance by overexpressing γ-glutamylcystein synthetase. *Plant Physiol.*, 121: 1169–1177.

Index